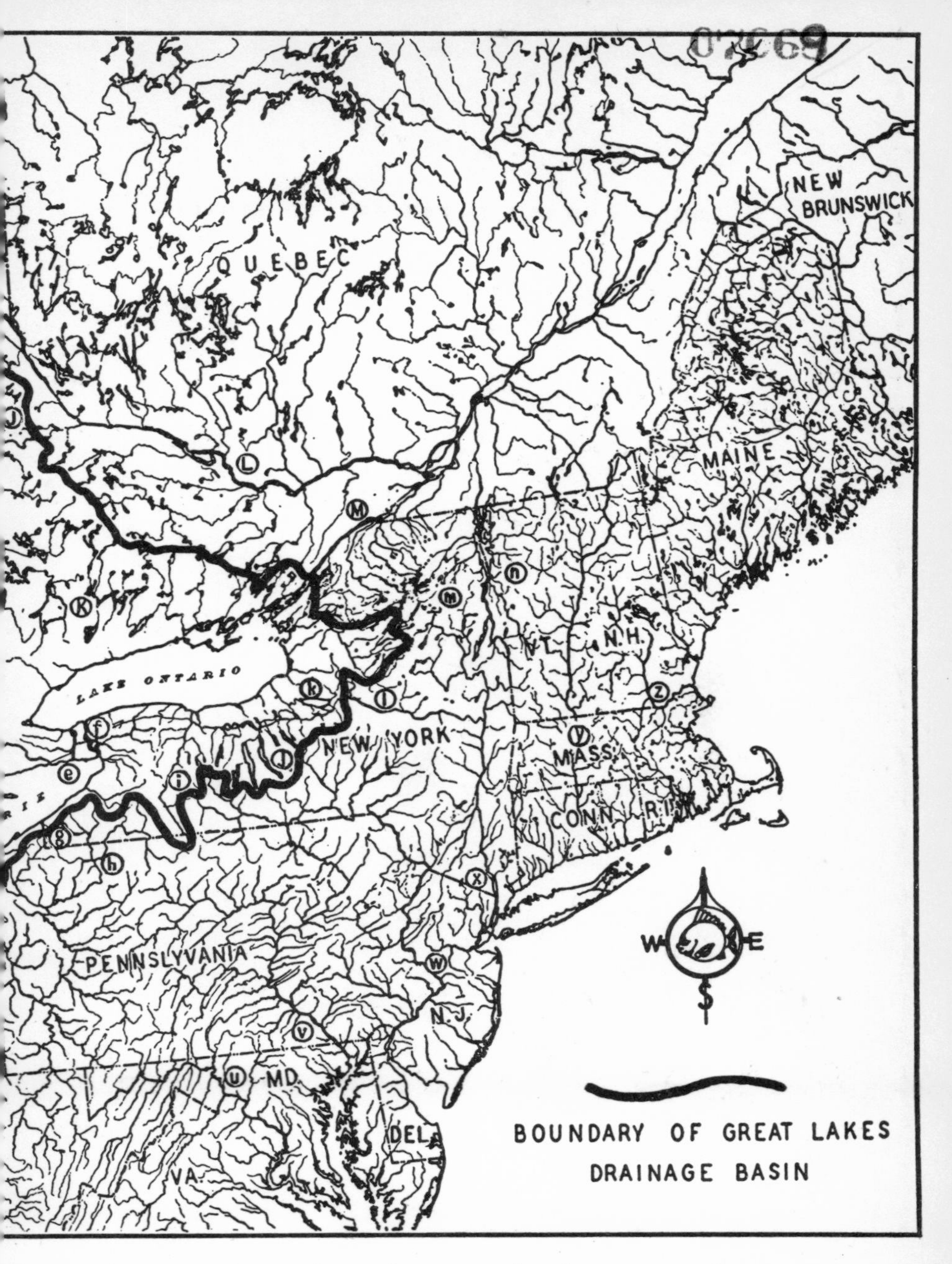

waters and localities that are frequently mentioned in the statement of ranges.

a. Maumee River, Ohio and Indiana
b. Detroit River
c. St. Clair River
d. Thames River, Ontario
e. Welland Canal, Ontario by-passing Niagara Falls
f. Niagara Falls
g. Chautauqua Lake, New York
h. Allegheny River

i. Genesee River, New York and Pennsylvania
j. Finger Lakes, New York
k. Oneida Lake, New York
l. Mohawk River, New York
m. Adirondack Mountains
n. Lake Champlain
o. Mississippi River
p. Wisconsin River
q. Missouri River

r. Illinois River, Illinois
s. Wabash River, Indiana and Illinois
t. Ohio River
u. Potomac River

Fishes of the Great Lakes Region

With a New Preface

by
Carl L. Hubbs
and
Karl F. Lagler

Ann Arbor
The University of Michigan Press

Fourth University of Michigan Press printing 1974
Originally published by Cranbrook Institute of Science,
Bloomfield Hills, Michigan
ISBN 0-472-08465-8
Library of Congress Catalog Card No. 58-7693
Published in the United States of America by
The University of Michigan Press and simultaneously
in Don Mills, Canada, by Longman Canada Limited
Manufactured in the United States of America

Preface, 1964

With the exception of a few mechanical corrections in the text and the new material in Preface, 1964, this book is a reprint of the 1958 edition. The history of editions prior to 1958 is given in the following Foreword. The 1958 edition was published as Bulletin 26 by Cranbrook Institute of Science. Publication rights have been assigned to the University of Michigan Press. We acknowledge with gratitude the kindness of Dr. Robert T. Hatt, director, and the Board of Trustees of Cranbrook Institute of Science, in assigning rights and in helping in many other ways.

Since the complete revision of this book for its 1958 edition several events have provided new information or background for interpretation. In this Preface are considerations of those events essential to the use of the 1964 printing.

One of the most significant of the events was, in 1960, Special Publication No. 2 of the American Fisheries Society, "A List of Common and Scientific Names of Fishes from the United States and Canada" by R. M. Bailey, *et al.* (2d ed.). This continues the Society's effort to encourage standardized use of both the common and technical names of fishes. Currently acceptable changes in fish names as given below, when not otherwise documented, are based on the foregoing list.

Additionally, taxonomic studies which have appeared have necessitated the revision of certain technical names of the fishes of the basin. These are given below and documented individually.

Finally, a few additions to the fish fauna of the Great Lakes through discovery and through intentional and inadvertent introductions are also given and documented below.

We have neither corrected any range extensions of fishes within or outside of the basin, nor have we added to the bibliography.

We are grateful to Drs. Reeve M. Bailey and Robert R. Miller, curators of fishes at the University of Michigan, for the continuation of their invaluable help to us.

Regarding fishes named after persons (patronyms), ichthyologists have sought uniformity through action by the International Commission of Zoological Nomenclature. Actions to date have failed to clarify the issue and optional procedure still prevails. Patronyms in this book follow old procedure (e.g., *bairdii*), whereas in the American Fisheries Society List the patronym is formed by adding *i* to the person's name (e.g., *bairdi*). At present, neither form given as an example can be termed correct, and the student should approach with understanding such differences among current published works in ichthyology.

In the following, we have called attention to certain problem patronymic spellings by showing what they become in the two contrasting procedures. At the present time we cannot endorse a change from our practice. The observations follow the sequence of the families, and within them, the species as they appear in order in the accounts of range and principal habitats.

The object of this prefatory material and of reprinting the 1958 edition of the book is to extend its usefulness at lowest possible cost to the users.

In *Entosphenus, lamottenii* may be given as *lamottei;* in spite of spelling, some uncertainty persists regarding the species name.

Lepisosteus productus (Cope) becomes *L. oculatus* (Winchell).

SHORTNOSE GAR, *Lepisosteus platostomus* Rafinesque, is added because of its recent discovery in Lake Winnebago, Wisconsin, by G. R. Priegel, Trans. Am. Fish. Soc., 92 (2): 178.

Amphiodon becomes *Hiodon* (p. 41).

Pomolobus becomes *Alosa* throughout; *pseudoharengus* (Wilson) becomes *Alosa pseudoharengus* (Wilson) and *P. chrysochloris* Rafinesque in footnote on p. 43 becomes *Alosa chrysochloris* (Rafinesque).

SALMON FAMILY, Salmonidae, with the WHITEFISH FAMILY, Coregonidae, and the GRAYLING FAMILY, Thymallidae, are combined in the single family Salmonidae, *fide* Norden, Jour. Fish. Res. Bd. Canada, 18 (5): 679-791.

EUROPEAN TROUT becomes BROWN TROUT, since for practical reasons in America, due to admixture of introduced lots of the species, subgroups have not appeared to retain their identity and the subspecies name, *fario,* that we have used does not seem tenable. Only the binomial *Salmo trutta* seems applicable.

Salmo clarkii lewisii (Girard) may also be given as *S. clarki lewisi* (Girard).

Salmo gairdnerii irideus Gibbons may also be *S. gairdneri irideus* Gibbons; Needham and Gard, Univ. Cal. Publ. Zool., 67, concluded that subspecies are not recognizable in the rainbow trout.

PINK SALMON, *Oncorhynchus gorbuscha* (Walbaum), is added as a result of its recent accidental introduction and subsequent establishment in Lake Superior, *fide* Schumacher and Eddy, Trans. Am. Fish. Soc., 89 (4): 371-373.

SHALLOWATER CISCO, *Coregonus artedii* LeSueur, becomes CISCO or LAKE HERRING; some authors prefer the name *Leucichthys* for this genus—e.g., Smith, Copeia, 1964 (1): 230.

NIPIGON TULLIBEE becomes NIPIGON CISCO.

GREAT LAKES BLOATER becomes, simply, BLOATER.

Osmerus mordax (Mitchill) becomes *Osmerus eperlanus mordax* (Mitchill), *fide* McAllister, Bull. Nat. Mus. Canada, 191 (Biol. Ser. 71), who also synonymized *O. sergeanti* (p. 57) with *o. e. mordax.*

In the genus *Catostomus, commersonnii* may also be spelled *commersoni.*

In *Erimyzon sucetta, kennerlii* may also be given as *kennerlyi,* if subspecies are valid in this fish; their validity has been challenged by Bailey, Winn, and Smith, 1954, Proc. Acad. Nat. Sci., Phila., 106: 100-164.

Moxostoma duquesnii (LeSueur) may also be given as *M. duquesnei* (LeSueur).

NORTHERN SHORTHEAD REDHORSE becomes NORTHERN REDHORSE.

NORTHERN CHUB becomes LAKE CHUB; LAKE NORTHERN CHUB becomes NORTHERN LAKE CHUB; and CREEK NORTHERN CHUB becomes WESTERN LAKE CHUB (if subspecies are recognizable.)

Doubt has been cast on the existence of subspecies in *Notropis atherinoides* by Bailey, 1956, *in* Iowa Fish and Fishing, Iowa State Cons. Comm., pp. 325-377.

Notropis cornutus does not contain recognizable subspecies according to Gilbert, Copeia, 1961 (2): 181-182, who also regarded *chrysocephalus* as a distinct species, the striped shiner, *Notropis chrysocephalus* (Rafinesque), with only the typical subspecies occurring in the Great Lakes.

Notropis roseus (Jordan) becomes N. texanus (Girard), *and N. roseus richardsoni* Hubbs and Greene becomes *N. texanus richardsoni* (Hubbs and Greene) if a northern subspecies is recognized.

Notropis deliciosus (Girard) becomes *N. stramineus* (Cope), and thus the SOUTHERN SAND SHINER becomes *N. stramineus subsp.* and the NORTHEASTERN SAND SHINER becomes *N. stramineus stramineus* (Cope), *fide* Suttkus, Copeia, 1954 (4): 307-318, if there are such subspecies; the subspecific question was reviewed by Bailey and Allum, 1962, Misc. Publ. Mus. Zool. Univ. Mich., 119.

Morphological differences for *Hybognathus nuchalis* and *H. placitus* (p. 85) have been desbribed by Bailey and Allum, *op. cit.*

Hybognathus nuchalis regia Girard becomes *H. n. regius* Girard to conform to Article 30 International Code of Zoological Nomenclature, 1961.

The LOACH FAMILY, Cobitidae, of Eurasia and Africa has become established in the Michigan watershed of Lake Huron, as evidenced by the 1958 and 1959 collection of the WEATHERFISH or DOJO, *Misgurnus anguillicandatus* (Cantor) by E. Schultz, Trans. Am. Fish. Soc., 89 (4): 376-377. The weatherfish is distinguished from the minnows (Cyprinidae) of the basin by its wormlike form, bottom-burrowing habit, and the possession of three pairs of prominent barbels on the upper lip.

Schilbeodes is replaced by *Noturus* throughout the madtoms.

Anguilla bostoniensis (LeSueur) becomes *A. rostrata* (LeSueur), following opinion 568 of the International Commission on Zoological Nomenclature.

Fundulus nottii dispar may also be given as *Fundulus notti dispar.*

COMMON GAMBUSIA becomes MOSQUITOFISH.

Gambusia affinis holbrookii Girard may also be given as *G. a. holbrooki* Girard

TROUTPERCH becomes TROUT PERCH for propriety.

PIRATEPERCH becomes PIRATE PERCH.

Morone becomes *Roccus, M. interrupta* (Gill) becomes *R. mississippiensis* (Jordan and Eigenmann), and *M. americana* becomes *R. americanus.*

YELLOW WALLEYE becomes WALLEYE.

BLUE WALLEYE becomes BLUE PIKE.

Hadropterus becomes *Percina* throughout, and, as a result, the spellings of species names change: *H. maculatus* to *P. maculata; H. phoxcephalus* to *P. phoxocephala.*

NORTHERN SAND DARTER becomes EASTERN SAND DARTER.

In *Etheostoma nigrum* the subspecies *olmstedi* is treated by some recent authors as a distinct species.

In *Micropterus dolomieu* the various forms are not regarded as subspecifically nameworthy by Bailey, 1956: *in* Iowa Fish and Fishing, Des Moines, Iowa State Cons. Comm., pp. 325-377.

FOURHORNED SCULPIN becomes FOURHORN SCULPIN, but differences of opinion exist on the systematic status of the form(s) we have called *Myxocephalus quadricornis thompsonii* (Girard); the latter term may also be spelled *thompsoni.*

In the genus *Cottus, bairdii* may be spelled *bairdi.*

In *Cottus,* the form *zopherus* (p. 118) is no longer regarded as a part of *C. bairdi,* e.g., C. R. Robins, Copeia, 1961 (2): 205-215.

In *Cottus cognatus,* McAllister and Lindsey, Bull. Nat. Mus. Canada, 172 (1961): 66-80, have concluded that subspecies are not valid.

Eucalia becomes *Culea* as shown by Whitley, Proc. Royal Zool. Soc., 1948-49 (1950): 44, for reasons summarized by Bailey and Allum, *op. cit.*

CARL L. HUBBS
La Jolla, California

KARL F. LAGLER
Ann Arbor, Michigan

Foreword

The present revision, although the first re-cast of Bulletin 26, is actually the third major refinement of works in the present series. The 1947 printing of Bulletin 26, under the title "Fishes of the Great Lakes Region" was an enlargement of our "Guide to the Fishes of the Great Lakes and Tributary Waters" (Bulletin 18, Cranbrook Institute of Science, 1941). In the years since the first appearance of Bulletin 26, much new information has come into being on the distribution of fishes that inhabit the Great Lakes drainage basin. We have tried to add this material, particularly to our statements of ranges. In the interval, too, there has been a tendency to change names of fishes, many in use for a long time. We have attempted to evaluate proposals for such change and to adopt those that seem warranted, because change by itself is not necessarily progress. We have not suppressed many of our earlier usages of subspecies, for clearly they have stimulated considerable taxonomic thought and need to stimulate even more before classification of the fishes of the Great Lakes region can be thought of as even nearly complete.

Serious students should not be discouraged from further study of the fishes of this area by the apparent completeness of parts of this book. In many ways the present work is only a beginning for intensive faunal studies of the region. Furthermore, the natural history of no one species is completely known, rather that of many is almost entirely unknown. We have attempted to point out some of the opportunities for investigation and are ready to help anyone seriously interested in pursuing such researches.

CARL L. HUBBS
KARL F. LAGLER

La Jolla, California
Ann Arbor, Michigan
August 1, 1957

Acknowledgments

In the preparation of this book we have had much help from many individuals and institutions. To all of these, particularly to our immediate colleagues and students who have helped us most, we are very grateful. Deserving of special mention are Reeve M. Bailey, Gerald P. Cooper, Raymond E. Johnson, Vianney Legendre, Robert R. Miller, George A. Moore, Edward C. Raney, W. B. Scott, Stanford H. Smith, Milton B. Trautman, Vadim D. Vladykov, James Zumberge, and, above all, Laura C. Hubbs.

Illustrations are from many sources. The line drawings illustrating characters were prepared by Janet Roemhild and Priscilla L. Woodhead. Joe Bender made the fish outlines for the family key. Chart I was composed by Stanford H. Smith.

Jack Van Coevering provided the original color photos of the brook and brown trouts. All of the other color plates are from cuts loaned by the Illinois Natural History Survey Division, Urbana. The publication of colored illustrations was aided initially by a generous grant from the Michigan Sportsman's Fund.

A. S. Hazzard, while director of the Institute for Fisheries Research, Michigan Department of Conservation, loaned several prints and provided the services of his technician, William Cristanelli, to make most of the originals reproduced as half-tones.

Vianney Legendre gave us the prints for Figures 32 and 109 and permission to adapt Fig. 24. Paul H. Eschmeyer and Reeve M. Bailey are responsible for the print used in Fig. 86.

Permission to copy several plates from the New York State Biological Survey Reports was granted by W. C. Senning. Similar kindnesses were extended by the Michigan Academy of Science, Arts and Letters, U. S. Fish and Wildlife Service, Illinois Natural History Survey Division, Boston Society of Natural History, Chicago Natural History Museum, and University of Michigan Museum of Zoology.

Several of the fish pictures reproduced as half-tones were made by the late Frank N. Blanchard as an aid in teaching. We deem it a great privilege to be able to put them to further educational use.

Individual acknowledgement as to source is made for each figure that is not our own.

Blanche Bell did the original library research on canal history and William Gordon provided range extensions from the data of the U. S. Fish and Wildlife Service.

The authorities of the Museum of Zoology of the University of Michigan have afforded us the use of the extensive records and collections of fishes housed there. The Department of Zoology, through its Chairman, G. R. LaRue, and, later, The School of Natural Resources, through Dean S. G. Fontanna, provided the time of several technical assistants and other facilities.

Data obtained by Lagler in summers 1945 through 1947 on the distribution of fishes on and about islands in the upper Great Lakes are included. The expeditions on which the collections were made were supported by a Rackham Research Grant and by the Michigan Department of Conservation.

For the present edition, Robert Wetzel was most helpful in Ann Arbor.

Because of the geographic separation of the authors and, further, repeated absences from the country during the preparation and publication of this revision, an inordinate amount of the labor and responsibility which should have been ours fell upon Robert T. Hatt, Director of Cranbrook Institute of Science, and upon his competent aides especially Margaret C. Fletcher. They have our deepest thanks.

TABLE OF CONTENTS

Brown Trout
Salmo trutta fario

LIST OF COLOR PLATES

PLATE 1.

Lake Sturgeon
Acipenser fulvescens

PLATE 2.

Bowfin (male)
Amia calva

PLATE 3.

Mooneye
Hiodon tergisus

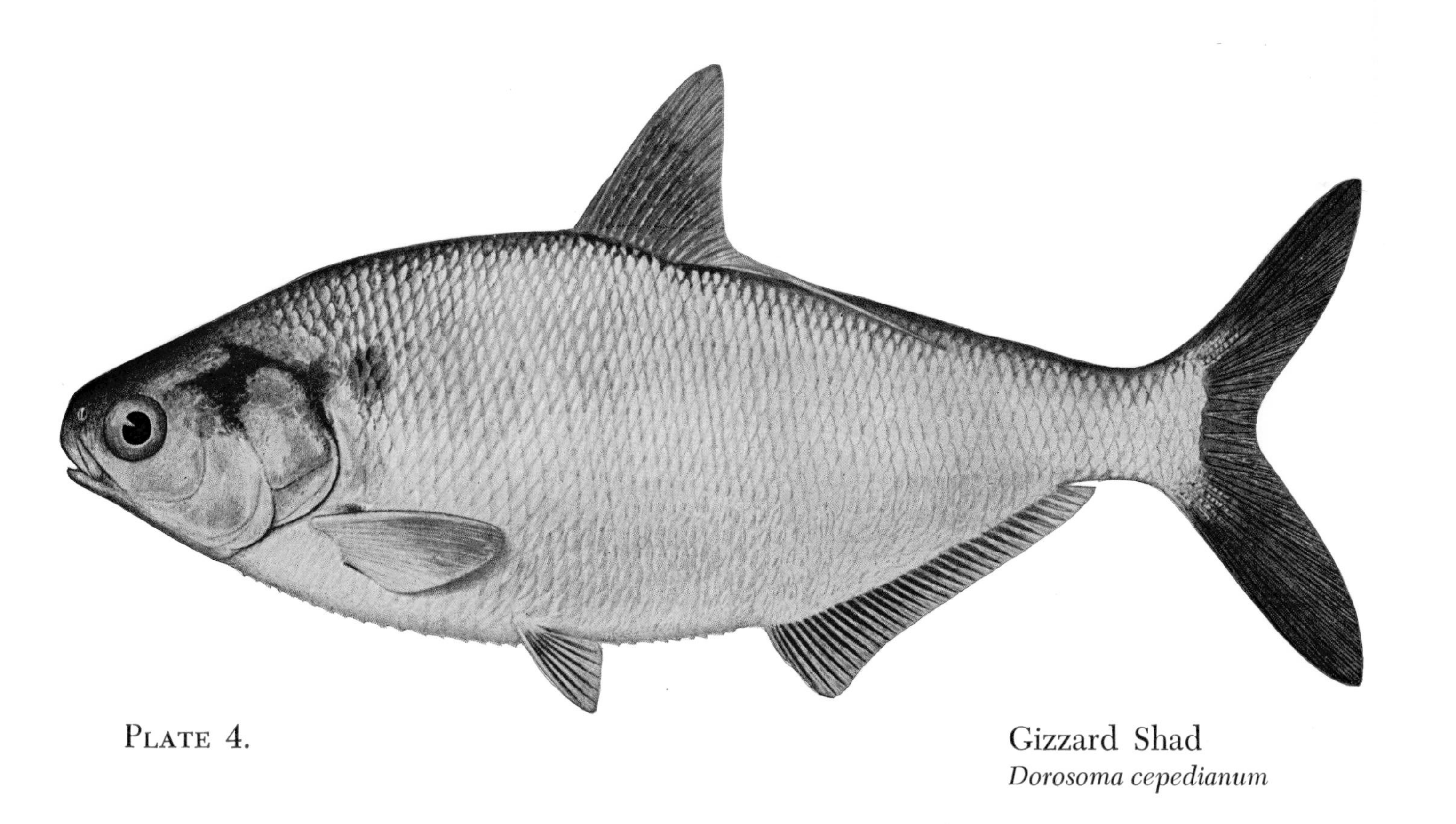

PLATE 4.

Gizzard Shad
Dorosoma cepedianum

PLATE 5. Brook Trout
Salvelinus fontinalis

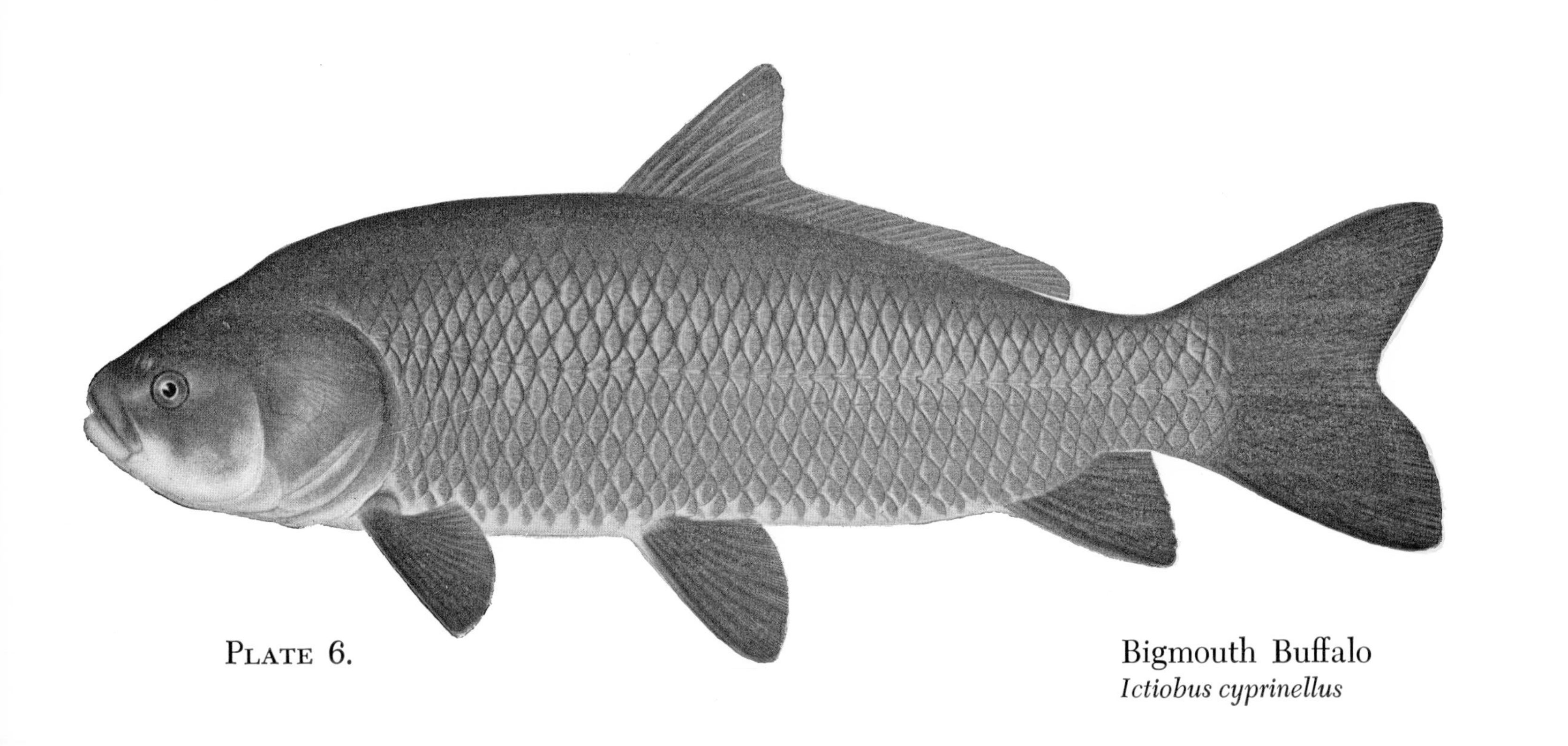

PLATE 6.

Bigmouth Buffalo
Ictiobus cyprinellus

PLATE 7.

Common White Sucker
Catostomus commersonnii commersonnii

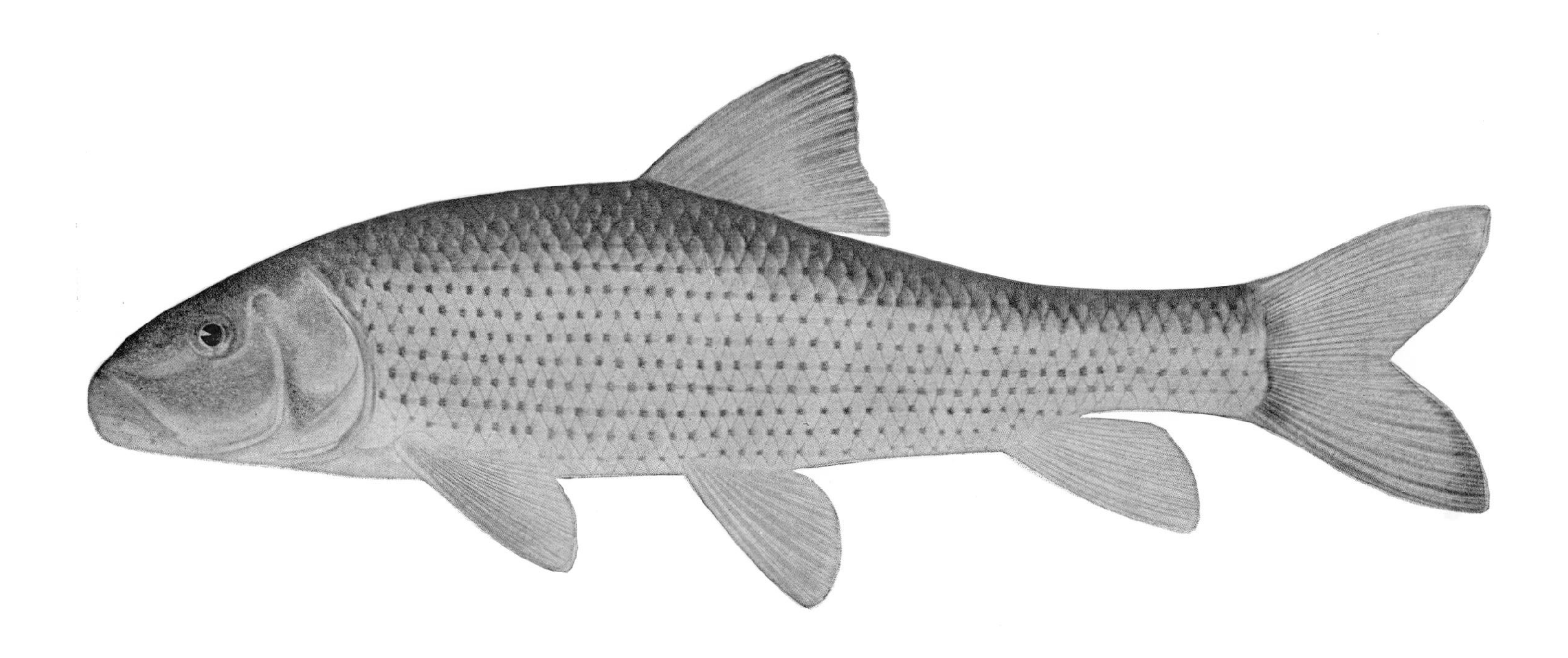

PLATE 8.

Spotted Sucker
Minytrema melanops

PLATE 9.

Western Creek Chubsucker
Erimyzon oblongus claviformis

PLATE 10.

Northern Shorthead Redhorse
Moxostoma macrolepidotum macrolepidotum

PLATE 11.

Northern Hog Sucker
Hypentelium nigricans

PLATE 12.

Carp

Cyprinus carpio

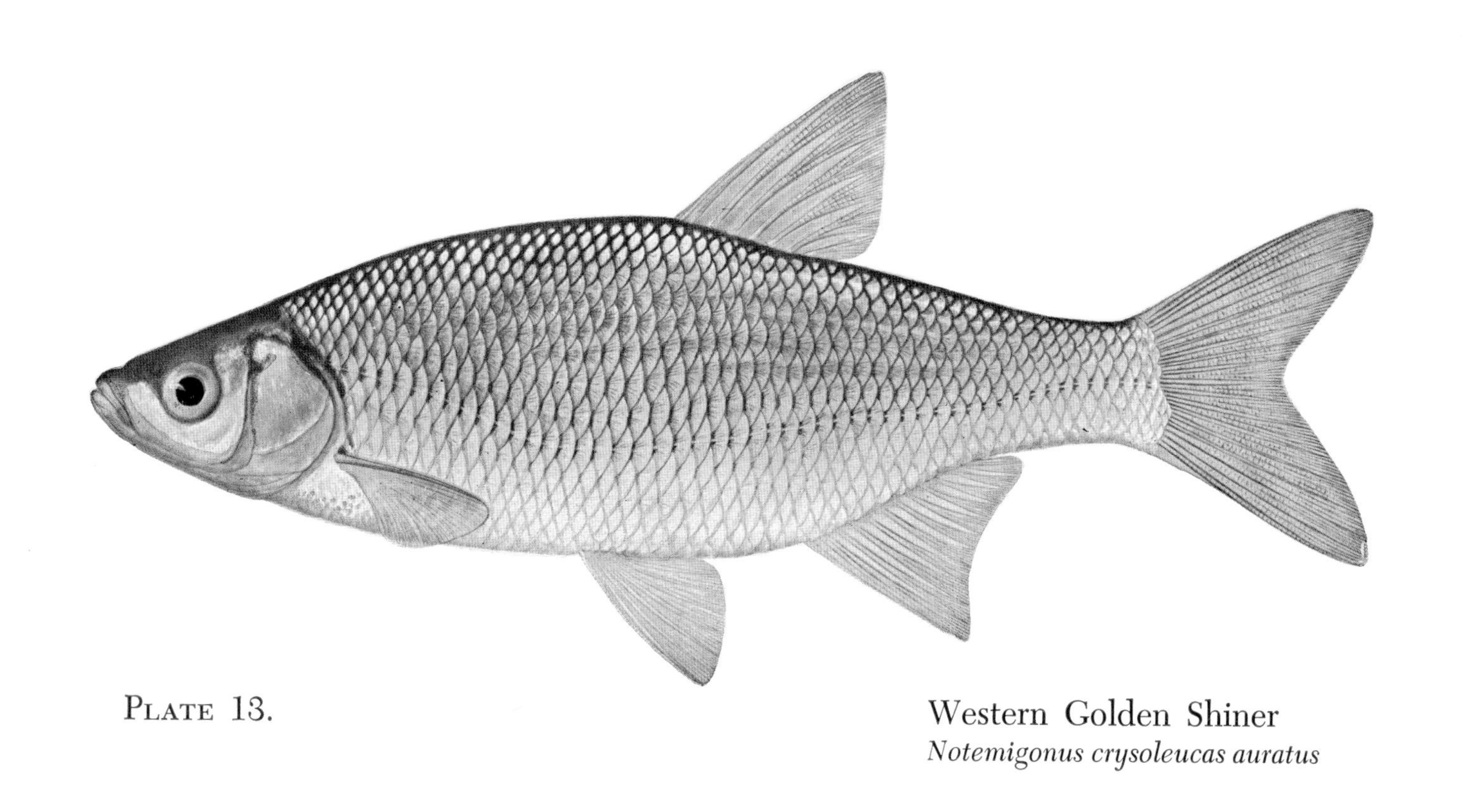

PLATE 13.

Western Golden Shiner
Notemigonus crysoleucas auratus

PLATE 14.

River Emerald Shiner
Notropis atherinoides atherinoides

PLATE 15.

Central Common Shiner
Notropis cornutus chrysocephalus

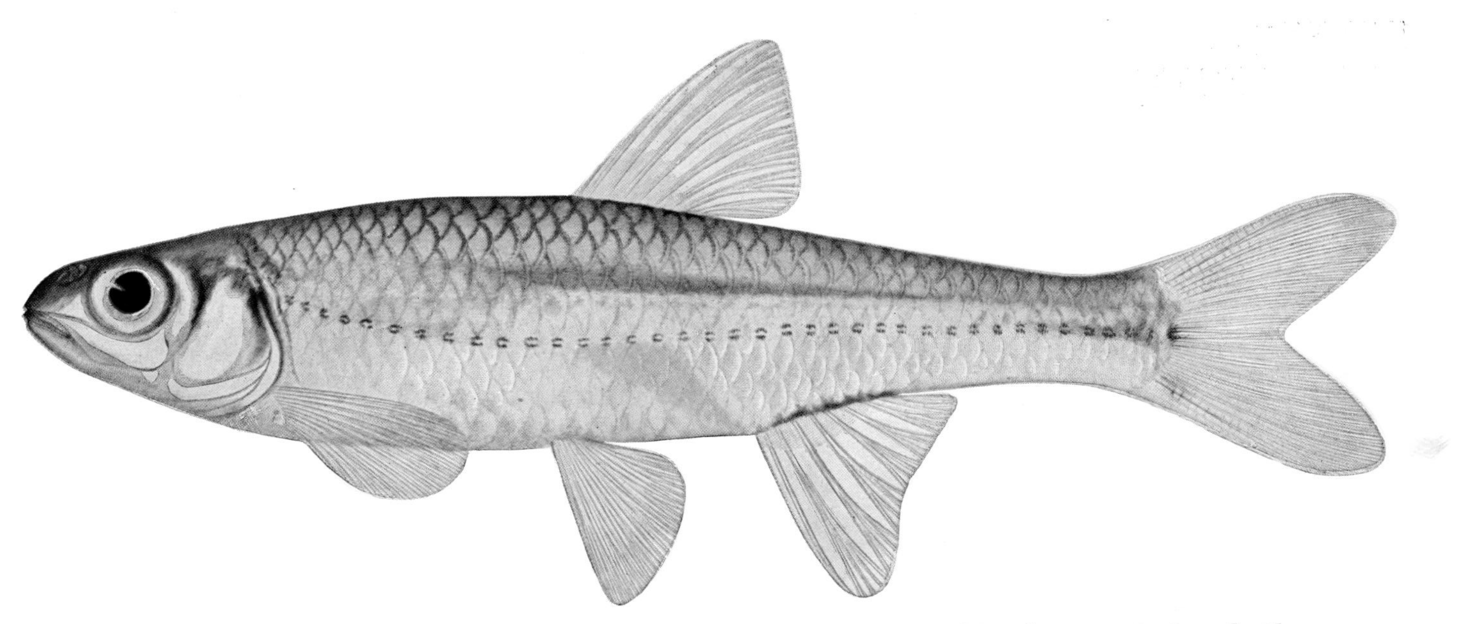

PLATE 16.

Northeastern Sand Shiner
Notropis deliciosus stramineus

Plate 17.

Suckermouth Minnow
Phenacobius mirabilis

Plate 18.

Northern Fathead Minnow
Pimephales promelas promelas

PLATE 19.

Central Stoneroller
Campostoma anomalum pullum

PLATE 20.

Northern Channel Catfish
Ictalurus punctatus punctatus

PLATE 21.

Northern Black Bullhead
Ictalurus melas melas

PLATE 22.

Northern Brown Bullhead
Ictalurus nebulosus nebulosus

PLATE 23.

Central Mudminnow
Umbra limi

PLATE 24.

Western Grass Pickerel
Esox americanus vermiculatus

PLATE 25.

American Eel
Anguilla bostoniensis

PLATE 26.

Western Banded Killfish
Fundulus diaphanus menona

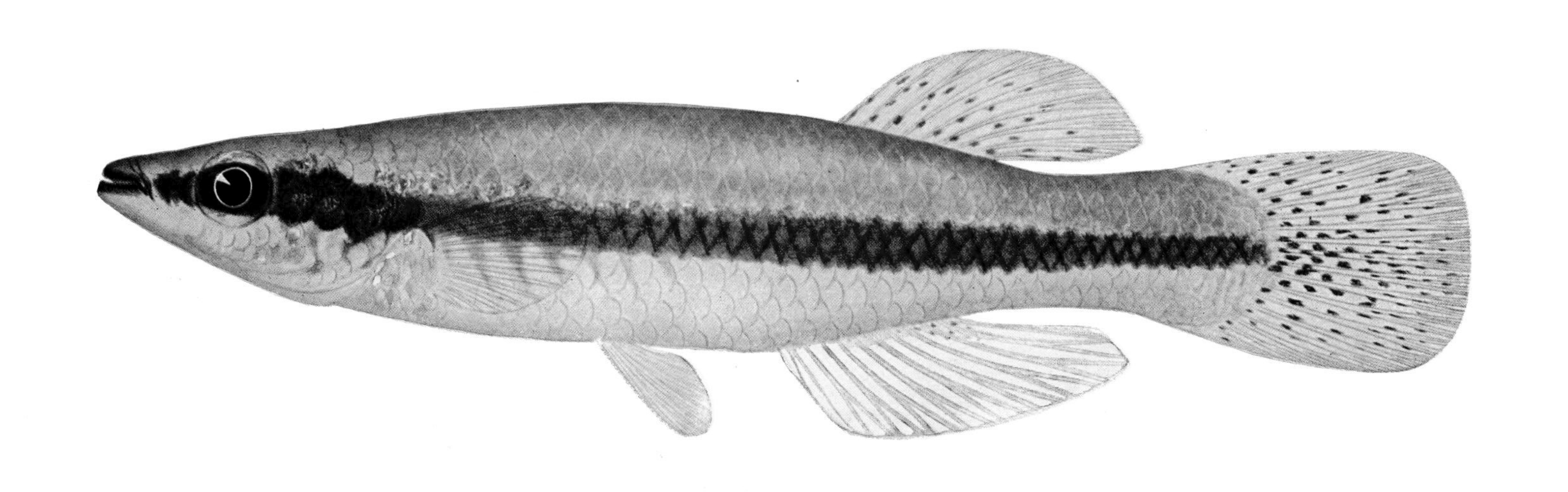

PLATE 27.

Blackstripe Topminnow
Fundulus notatus

PLATE 28.

Yellow Perch
Perca flavescens

PLATE 29.

Yellow Walleye
Stizostedion vitreum vitreum

PLATE 30.

Blackside Darter
Hadropterus maculatus

Plate 31.

Northern Logperch

Percina caprodes semifasciata

PLATE 32.

Iowa Darter
Etheostoma exile

PLATE 33.

Striped Fantail
Etheostoma flabellare lineolatus

PLATE 34.

Northern Greenside Darter
Etheostoma blennioides blennioides

PLATE 35.

Northern Smallmouth Bass
Micropterus dolomieui dolomieui

PLATE 36.

Northern Largemouth Bass
Micropterus salmoides salmoides

PLATE 37.

Green Sunfish
Lepomis cyanellus

PLATE 38.

Common Bluegill
Lepomis macrochirus macrochirus

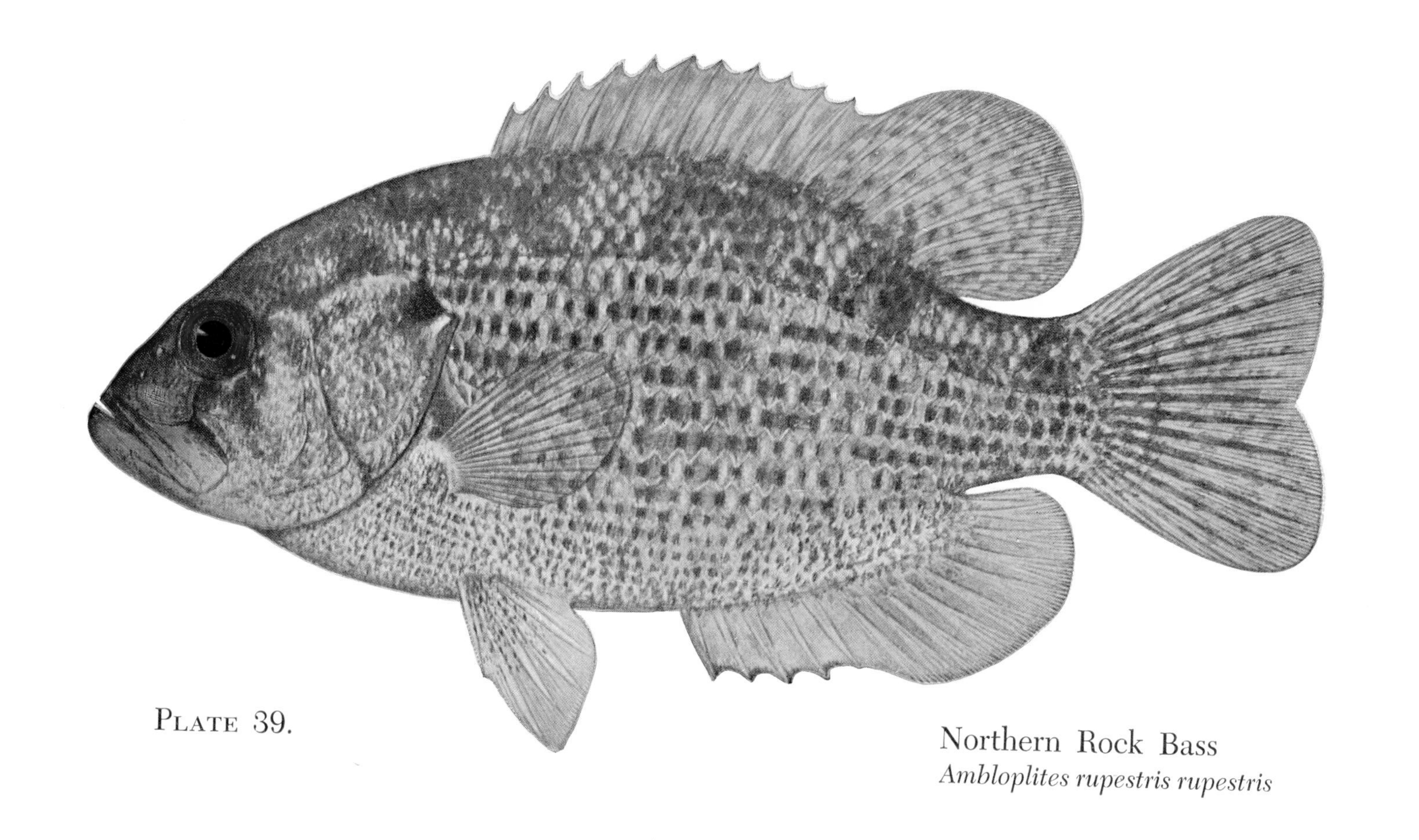

PLATE 39.

Northern Rock Bass
Ambloplites rupestris rupestris

PLATE 40.

White Crappie
Pomoxis annularis

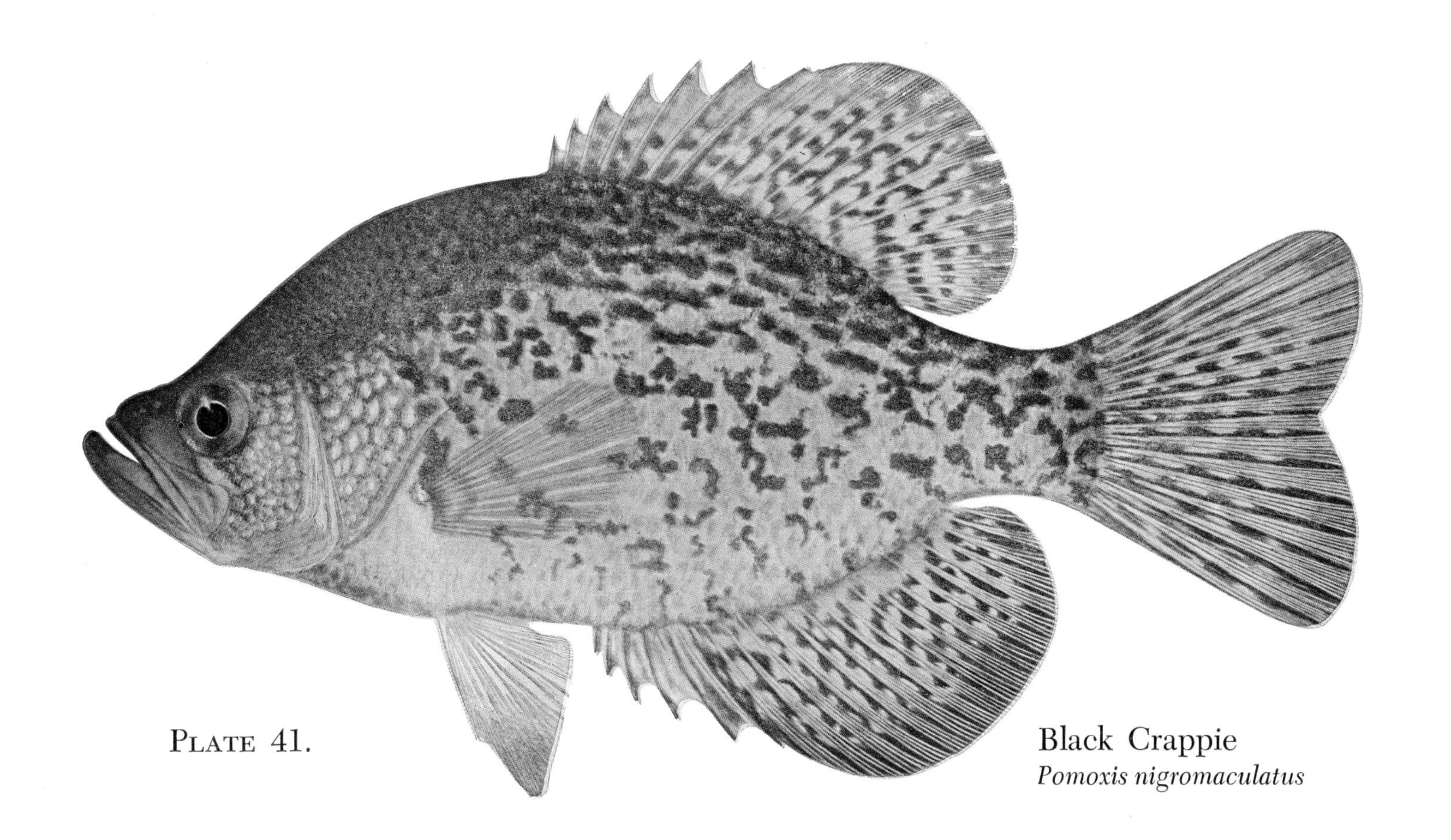

PLATE 41.

Black Crappie
Pomoxis nigromaculatus

PLATE 42. Northern Brook Silverside
Labidesthes sicculus sicculus

PLATE 43.

Freshwater Drum
Aplodinotus grunniens

PLATE 44.

Brook Stickleback
Eucalia inconstans

Fishes of the Great Lakes Region

By Carl L. Hubbs
Scripps Institution of Oceanography, University of California
and
Karl F. Lagler
Department of Fisheries, University of Michigan

INTRODUCTION

Early in his career the senior author, a student of David Starr Jordan, arrived at a conclusion and a motto—"Where there is water, there are fishes, and where there are fishes they can be caught." Few regions of the world are so richly and so diversely endowed with fresh waters as is the basin of the incomparable Great Lakes, and in few regions of the world can such a varied representation of freshwater fishes be caught. It is therefore only natural that much interest has developed in the study of the fishes of the Great Lakes and tributary waters.

The richness and variety of the Great Lakes fish fauna stems not only from the great diversity of the abundant waters—grading from warm to very cold, from stagnant to swift and turbulent, from small brooks and tea-colored bog ponds to great rivers and profoundly deep inland seas–but also from the diverse origin of the fishes. Some of the species are relicts of arctic and subarctic waters, forced south by the continental ice sheets of the Pleistocene Age and now trapped in the Great Lakes and some of the inland waters. A large number entered the basin from the South and from the East, as the ice cap retreated. A few came in from the sea through the St. Lawrence River. There is a large representation of the rich and varied fish fauna of eastern North America, including such ancient kinds—"living fossils"—as the paddlefish, gars, bowfin, mooneye, troutperch, and pirateperch.

In parts of the basin the waters and the fish life have been modified by the activities of man, and the fish fauna has been increased by purposeful introductions and by passage through man-made canals (the destructive sea lamprey is one of the canal immigrants). Many of the waters have deteriorated through the march of civilization, and the fish life has correspondingly suffered. Various types of pollution ranging from inorganic silt through organic and industrial wastes have had an evil effect. In some waters, like Lake Erie, the more valuable and highly prized fishes have been largely replaced by coarser kinds. Some stretches of water have become almost open sewers, all but devoid of fish life. A few species have been extirpated in the basin. But over much of the Great Lakes area the waters and their finny inhabitants remain in a semi-virgin condition, ready for study. The maintenance and restoration of natural conditions is a challenge to the human population of the basin.

THE WATERS OF THE GREAT LAKES REGION AND THEIR FISH ASSOCIATIONS

Geographically, the Great Lakes lie midway between the Equator and the North Pole. The watershed straddles both 45° N. Latitude and 85° W. Longitude. The natural drainage basin now covers 287,770 square miles (Table I). The Great Lakes themselves occupy about 33 per cent of this total area, or 94,700 square miles and have an aggregate shoreline of more than 4,000 miles. The influence of the lakes is important in determining the composition of the fauna of the region and the conditions under which the animals live. For example, the lakes temper the climate in their immediate environs and strongly affect precipitation.

TABLE I

PHYSICAL FEATURES OF THE GREAT LAKES*

Feature	*Lakes*				
	Superior	*Michigan*	*Huron*	*Erie*	*Ontario*
Surface area, sq. miles	31,820	22,400	23,010	9,930	7,540
Natural drainage area, sq. miles	80,000†	67,860	72,620	32,490	800
Dissolved solids, ppm.	60	118	108	133	134
Approximate August temperature, °F.	55	67	67	74	70
1900 - 1954 Mean datum plane, feet (Elevation above sea level)	602	580	580	572	246
Maximum depth, feet	1333	924	750	210	778
Mean depth, feet	487	276	195	58	264

* Based in part on U.S. Lake Survey, Great Lakes Pilot, 1956.

† Plus 5545 square miles added in the Ogoki diversion, 1943, and 1550 square miles in the Long Lac diversion, 1938, both in Ontario.

In depth, the Great Lakes range from deep Lake Superior with a maximum of 1333 feet (average, 487 feet, with the deepest part about 721 feet below sea level) to shallow Lake Erie. The mean depth of Lake Erie is only 58 feet and the maximum is 210 feet in a relatively small hole. Mean August temperatures for these two bodies of water are the most diverse—55° and 74° F. respectively.

Innumerable inland lakes grace the watershed. The total perhaps exceeds 50,000—Michigan alone claims in excess of 11,000. These lakes, like the thousands of miles of streams that flow through and among them, are classifiable in many ways. One major division would be into those that could or do support trout ("cold waters") and those that do not ("warm waters"). All of the Great Lakes, excepting Lake Erie, are inhabited by trout and even the deep hole of Lake Erie supports coldwater coregonids. Many inland lakes and streams north of Parallel 43°, and a few south of it, are cold.

Another grouping of waters of the Great Lakes drainage basin can be made on the basis of their dissolved contents, which in turn are related to the strength of the food chains on which fish populations depend. Waters entering the basin from the north, as well as from Wisconsin, and those in the western half of the Upper Peninsula of Michigan, are soft and of low fertility. They course over ancient, Precambrian rocks and the result of their outfall is seen, for example, in Lake Superior's low content of dissolved solids, 60 parts per million (Table I). In the other Great Lakes, dissolved solids average approximately twice as high. Reflecting an interaction of chemical content of the water, mean depth, temperature, species content, fishing effort and efficiency, and other factors, the recorded commercial catch of fishes from Lake Erie (in recent years) is two to five times that for any other of the Great Lakes.

Inland lakes of the Great Lakes region (as well as bays and inshore waters of the big lakes) become ice-covered in most winters. However, not all of them stratify thermally in midsummer. By thermal stratification is meant the temperature layering that characterizes lakes of the temperate zones in midsummer. This summer layering is into an upper warmwater zone (the epilimnion), a layer of rapid temperature drop (the thermocline), and a deeper, coolwater zone of even temperature (the hypolimnion, with a temperature of 38° to 40° F.). Many of the lakes that stratify thermally in the summer are inhabited by coldwater fishes. Most lakes that are shallower than 20 feet do not stratify. Some such lakes are subject, when ice-covered for prolonged periods, to winterkill of fish. These are not inhabited ordinarily by salmonids or coregonids, rather they have various complexes of warmwater species including sunfishes. Some shallow waters are bog ponds in which the water is tea-colored and often acid.

Bog waters and the depths of the Great Lakes exhibit two extremes of habitat and fish inhabitants. Typically bog ponds are lakes nearing extinction (dystrophic lakes). The depths of Lake Superior, in contrast, represent waters of the region that are about as "new" or "virgin" as waters can be (as in oligotrophic lakes). They exemplify conditions very nearly like those in existence not long after the disappearance of the last glacier, 11,000 to 13,000 years ago. A typical bog fish association includes the northern redbelly dace, the mudminnow, and the brook stickleback. The deepwater community in the Great Lakes proper comprises the lake trout, burbot, deepwater sculpin, and deepwater ciscoes ("chubs").

The shore fauna of the Great Lakes and a few of the largest inland lakes of the basin is fairly distinctive. The species associations depend on location. Common associates in the surge zone are the longnose dace and Great Lakes mottled sculpin. Relatively quiet waters along exposed shores are frequented by the young of the yellow perch and by shiners (lake emerald, mimic, and sand). At night, particularly in the Upper Lakes, the troutperch moves shoreward. Where there is cover, the rock bass is to be found, and, in many pockets of abundance, the smallmouth bass. In Lake Ontario, and expectedly now in the Upper Lakes, young of the alewife frequent shore waters. In cold Lake Superior the young of the smelt or the whitefishes and the ninespine stickleback come into the shallows.

In weedy inland lakes, or in well-vegetated coves of otherwise open lakes, such fishes as the bluntnose minnow, the blacknose and blackchin shiners, and bullheads are apt to be found. Along the exposed sandy shores of inland lakes the mimic and the sand shiners are likely to be encountered. Where the bottom is predominantly soft, largemouth bass, northern pike, and the Iowa darter are to be expected, but when it is firm, the smallmouth and the yellow walleye abound.

Coldwater streams contain stream trouts, both naturally produced and stocked. Commonly associated with them are the creek chub, white sucker, common shiner, and slimy sculpin. On the riffles of such waterways are darters and the blacknose dace. Also there, in the swiftest currents, is the longnose dace, which is also found in the surf of the Great Lakes shores. In quiet backwaters with silty bottom, the brook stickleback is present.

Warmwater streams are often occupied by the largemouth bass, if the watercourse is large enough and quiet enough. If the current is strong, the smallmouth bass is to be expected along with, in large rivers, the yellow walleye. Common associates, besides crayfishes, of the foregoing stream occupants are the river chub, various darters, rock bass, and sunfishes.

In warm impoundments, and in Lake Erie, the crappies (both black and white species) do very well. The introduced carp is also often to be found in such waters.

ZOOGEOGRAPHY

The waters comprising the Great Lakes and tributaries of today (for map see endpapers inside covers) contain an interesting and extensive fish fauna that includes representatives of the most important families of North American fresh waters (Table II). Among these families are several that include ancient relicts—"living fossils"—unknown from other continents, namely, the families of the gars (Lepisosteidae), bowfin (Amiidae), mooneyes (Hiodontidae), troutperches (Percopsidae), and pirateperch (Aphredoderidae). Other ancient relicts are the paddlefish (Polyodontidae and the mudminnows (Umbridae), each of which belongs to a group with one representative in the Old World. Two other families, each important in the fauna of the Great Lakes region, are exclusively North American. These are the families of the North American catfishes (Ictaluridae) and the sunfishes (Centrarchidae). In the perch family (Percidae), the darters (subfamily Etheostomatinae) comprise a speciose North American group rather well represented in the Great Lakes fauna.

Two Great Lakes fishes, the brook silverside and the freshwater drum, arose from distinctively North American branches of marine families (respectively Atherinidae and Sciaenidae). The

TABLE II

TALLIES OF GROUPS AND FORMS OF GREAT LAKES FISHES BY FAMILIES

Arranged in order of number of species and genera

Family	*Genera*	*Species*	*Total Forms*
Minnows (Cyprinidae)	17	50	65
Suckers (Catostomidae)	8	19	23
Perches (Percidae)	6	19	24
Whitefishes (Coregonidae)	3	14	45
Sunfishes (Centrarchidae)	5	12	12
N. Am. Catfishes (Ictaluridae)	4	12	12
Salmons (Salmonidae)	3	7	10
Lampreys (Petromyzontidae)	3	5	5
Sculpins (Cottidae)	2	4	5
Pikes (Esocidae)	1	4	4
Herrings (Clupeidae)	3	3	3
Sticklebacks (Gasterosteidae)	3	3	3
Killifishes (Cyprinodontidae)	1	3	4
Gars (Lepisosteidae)	1	2	2
Basses (Serranidae)	1	2	2
Paddlefishes (Polyodontidae)	1	1	1
Sturgeons (Acipenseridae)	1	1	1
Bowfins (Amiidae)	1	1	1
Mooneyes (Hiodontidae)	1	1	1
Graylings (Thymallidae)	1	1	1
Smelts (Osmeridae)	1	1	1
Mudminnows (Umbridae)	1	1	1
Freshwater Eels (Anguillidae)	1	1	1
Cods (Gadidae)	1	1	1
Troutperches (Percopsidae)	1	1	1
Pirateperches (Aphredoderidae)	1	1	1
Silversides (Atherinidae)	1	1	1
Drums (Sciaenidae)	1	1	1

three species of killifishes (Cyprinodontidae) in the Great Lakes fauna belong in one of the North American divisions of the family. Among the lampreys (Petromyzontidae) of the Great Lakes, three of the five species belong in the North American genus *Ichthyomyzon,* which is probably another ancient relict.

The other families of Great Lakes fishes, like the Petromyzontidae, are members of the great Holarctic fauna, which is widespread around the world in northern regions. In fact, most of these groups probably originated in southeastern Asia and emigrated to North America during Tertiary time, so that the American genera and species are of relatively recent origin. The minnows (Cyprinidae) have multiplied prodigiously in the New World, and many live in the Great Lakes basin. The suckers (Catostomidae) have also done well, though they have almost vanished in Asia. The whitefishes (Coregonidae) have prospered particularly well in the Great Lakes. The ten other families native in the Great Lakes contain only one to four species each in the area covered by this book. (See Table II.)

POSTGLACIAL REDISPERSAL

Since the entire drainage area of the Great Lakes was covered by ice during the several advances of the continental ice sheet, including the last major advance (the Wisconsin), the rich native fish fauna now present must have reinvaded the basin from ice-free refuges. A few are presumed to have entered the basin or have extended their ranges much more recently, through canals and through stocking.

TABLE III

SUMMARY OF DEVELOPMENTAL AND DRAINAGE HISTORY OF THE GREAT LAKES*

Stage	*Approximate Location of Stage*	*Drainage Connection*
PREGLACIAL DRAINAGE	St. Lawrence drainage basin	Laurentian River—Atlantic Ocean
GLACIAL LAKES		
L. Duluth	Western L. Superior	Mississippi Valley
L. Chicago	L. Michigan	Mississippi Valley
L. Maumee	Western L. Erie	Mississippi Valley
L. Saginaw	Saginaw Bay of L. Huron	Mississippi Valley
L. Arkona	Southern L. Huron and L. Erie	Mississippi Valley
L. Whittlesey	L. Erie region	Mississippi Valley
L. Wayne	L. Erie to Finger Lakes region	Atlantic O. *via* Hudson R.
L. Warren	L. Erie and southern L. Huron regions	Mississippi Valley
L. Lundy	L. Erie region	Atlantic O. *via* Hudson R.
L. Iroquois	L. Ontario basin	Atlantic O. *via* Hudson R.
L. Algonquin	Upper Great Lakes basins	Mississippi Valley, and Atlantic O. *via* St. Lawrence R.
Nipissing Great Lakes	Great Lakes basins	Atlantic O. *via* St. Lawrence R.
PRESENT GREAT LAKES	Great Lakes basins	Atlantic O. *via* St. Lawrence R., natural; other connections *via* canals, see Table IV.

* Data mostly from: "The Pleistocene of Indiana and Michigan and the History of the Great Lakes" by Frank Leverett and Frank B. Taylor, U. S. G. S. Monogr., Vol. 53, 1915, 529 pp.

This table was prepared with the help of Professors I. D. Scott, George M. Stauley, and James H. Zumberge; it is over-simplified. The relations of the various Glacial Lakes are not completely known.

As the continental ice cap retreated, bodies of meltwater formed along its southern edge, dammed by the ice front. These ice-marginal lakes grew larger with the disappearance of the glacier and eventually spread over a combined area greater than that of the present Great Lakes. As the Glacial Lakes grew and land levels changed, outlets were attained in several directions. These overflows allowed fishes to re-enter the Great Lakes region. Table III is a condensed summary of the ancestral history of the Great Lakes.

The Postglacial centers of origin for the present fauna were principally those of the Northwest, the upper Mississippi Valley, and the middle Atlantic Costal Plain. At various Glacial stages over long periods of time, broad drainage connections were afforded by run-off from the melting ice over present drainage divides between the Great Lakes watershed and the surrounding ones (Fig. 1 and Table III). The principal routes for fish migration thus afforded were: (1) from the Northwest, by early passage through the connections with the upper Mississippi basin (and possibly from glacial Lake Agassiz of the Winnipeg region to the Hudson Bay drainage and thence southward to the Great Lakes); (2) from the South through various combinations of glacial outlets to the Mississippi and Ohio river valleys (the fish using in part the same connections, as the Chicago Outlet, that had earlier been used by the species from the Northwest); (3) from the East through several effluents, including those in the present St. Lawrence and Mohawk-Hudson river channels. Because of these drainage cross-connections fishes repopulated the area as the ice left, and spread through it as environmental conditions and water connections became suitable. Ecological barriers and changing conditions have served not only to restrict or halt the spread of certain forms but also to extirpate the populations of others that had penetrated into the region. For certain additional fishes, dispersal has not yet run its fullest possible extent. The water routes through the Great Lakes basin also afforded means by which central and northwestern fishes entered the Atlantic Costal Plain fauna. These conclusions regarding the principal origins of the contemporary Great Lakes fishes stem from three sets of data: (1) distributional patterns of the various forms, as given subsequently in the statements of ranges under each species; (2) the present centers of abundance; and (3) the available avenues of entry shown by Glacial geologists to have existed during recession of the continental ice cap.

Entry and dispersal routes cannot be postulated with assurance for all forms. Certain ones, however, almost surely utilized specific courses; examples follow. *Entry from the Northwest*: lake trout, whitefishes, ciscoes, troutperch and three sculpins. *Entry from the South:* gars, bowfin, spotted sucker, redhorses, many shiners, fathead minnow, and western grass pickerel. *Entry from the East*: cutlips minnow, fallfish, and bridled shiner. *Probable entry from both East and West*: yellow perch, pumpkinseed, and the eastern and western pairs of subspecies of the blacknose dace, banded killifish, and Johnny darter.

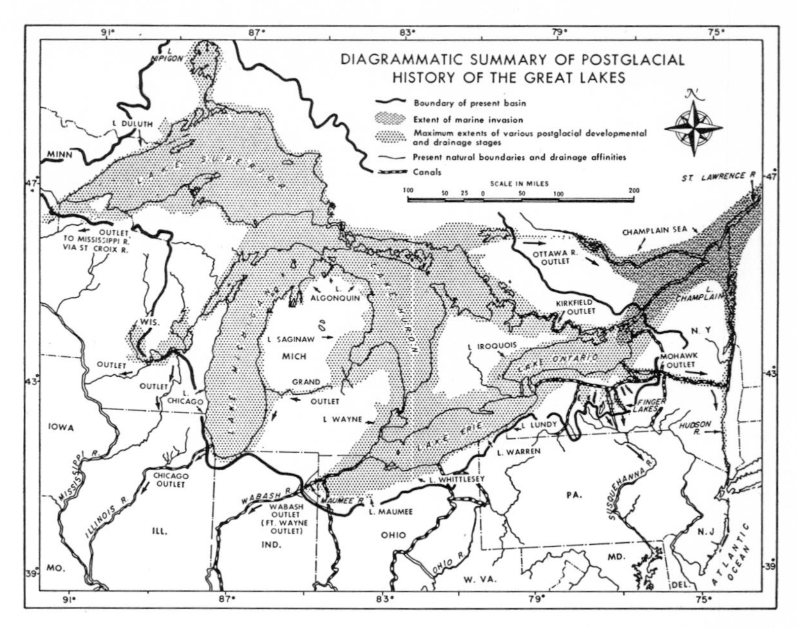

Fig. 1. Diagrammatic summary of the Postgiacial history of the Great Lakes showing the many available water routes to other drainages. Several of these waterways to the West, South and East were avenues of entry and dispersal for fishes during and after recession of ice from the region. Outlets during Glacial times are underscored: general locations of Glacial lake stages are indicated by arrows. The outlets were never all co-existent and the maximum extent of the lakes as shown was never attained simultaneously in all parts of the basin.

With the exception of the St. Lawrence River, all major natural drainages of Glacial times from the Great Lakes into the Mississippi River and into the Atlantic Ocean have almost ceased to exist. However, there is at least one persisting natural water route to the drainage of Hudson Bay. Shallow Summit Lake straddles the divide and contributes water back to Lake Nipigon (via the Ombabika River) and to Hudson Bay (via the Ogoki–Albany system).

In the early days of the white man in the Great Lakes region there were still connectives, at least intermittent, to the Mississippi. The Jesuits reported that in the freshets of a rainy spring a canoe might be paddled across the mile of low divide between the Lake Michigan drainage of the Chicago River and the Des Plaines River of the Mississippi basin. In 1914, flood water was observed in the region of Fort Wayne, Indiana, connecting the Maumee River (Lake Erie tributary) with the Wabash River of the Mississippi system. The connection was in the old channel of outlet of glacial Lake Maumee. As late as 1820, in seasonally high waters, army barges were floated down the Fox River, from the Wisconsin River to Green Bay, Lake Michigan. Subsequently, at these locations (and elsewhere), canals were dug that enlarged or restored the natural connections, or made new ones (Fig. 1

and Table IV). At least some of these waterways, built since the 1820's, have served to upset the natural fish geography, as is indicated for several species in the following text. It is possible that certain of these canals, some of which are no longer in use, have produced temporary faunal changes or have brought about changes that as yet have escaped notice. Niagara Falls would have remained a complete natural barrier separating Lake Ontario from the other Great Lakes were it not for such connections as the Welland Canal and the Trent Canal. The St. Lawrence Seaway enhances the connection of the entire Great Lakes basin with the Atlantic Ocean. Examples of the zoogeographic effects of canals are the following:

(1) The sea lamprey originally had not penetrated the system above Niagara Falls, but now abounds throughout the Great Lakes, presumably as a result of passage through the Welland Canal.

(2) The alewife was unknown from the Upper Lakes prior to the opening of the Welland Canal but has recently completed its spread to all of them.

(3) The Chicago Drainage Canal has provided a modern link with the Mississippi watershed by which several species may have penetrated into Lake Michigan.

(4) The recent appearance of the fallfish in northern Lake Superior tributaries seems to be the result of new drainage connectives established for hydroelectric purposes with streams flowing into Hudson Bay.

In recent years, diversion of two segments of the Hudson Bay drainage enlarged the watershed of Lake Superior by 7095 square miles through artificial stream piracy. Minor faunal additions may have resulted. One of these diversions brings into Lake Nipigon the flow from a major part of the Ogoki River system (originally Albany River—Hudson Bay watershed). The other, just east of Lake Nipigon, reverses by a dam the outlet of Long Lake (Long Lac) from the Kenogami River (of the Albany system) to make it tributary to Lake Superior *via* the Aguasabon River.

Several kinds of fish have been introduced by stocking into the Great Lakes and tributary waters. These introductions have met with varying degrees of success. Exotics which have become at least locally established are: (1) American smelt from the Atlantic Coast (native in the Lake Ontario basin); (2) brown trout from Europe; (3) rainbow trout from the West Coast; (4) carp from Asia *via* Europe; (5) goldfish from Asia; (6) redear sunfish from the Mississippi Valley; and (7) gambusia, the "mosquitofish," from the Gulf of Mexico drainage. Of temporary establishment were: (1) landlocked Atlantic salmon from Maine; (2) Yellowstone cutthroat from the West; and (3) chinook salmon from the Pacific Coast. The American eel was stocked in Michigan but has, of course, failed to reproduce and is either extinct or nearly so;

TABLE IV

SUMMARY OF CANALS AND DIVERSIONS OF POSSIBLE SIGNIFICANCE IN THE FISH GEOGRAPHY OF THE GREAT LAKES REGION*

Canal or Diversion	*Watershed Connections*	*Year of Completion*
Fox River (Portage) Canal	Fox River (L. Mich.)—Miss. R.	About 1840
Chicago Drainage Canal	Southern L. Michigan—Mississippi R.	1840
Wabash-Erie Canal	Western L. Erie—Ohio R.	1836
Miami-Erie Canal	Western L. Erie—Ohio R.	1830
Ohio-Erie Canal	Middle L. Erie—Ohio R.	1832
Ohio-Erie Canal Branches		
Pennsylvania-Ohio (Mahoning)	Ohio-Erie Canal—upper Ohio R.	1840-50
Sandy and Beaver	Ohio-Erie Canal—upper Ohio R.	1850 (?)
Erie (Barge) Canal	L. Erie—upper Hudson R.	1825
Erie Canal Branches		
Champlain	Hudson R.—L. Champlain	1819
Chemung	Seneca L.—Susquehanna R.	1833
Chenango	Mohawk R.—Susquehanna R.	1838
Cayuga-Seneca	Cayuga L. and Seneca L.—Erie Canal	1829
Black R.	L. Ontario—Mohawk R.	1840
Oswego	L. Ontario—Mohawk R.	1828
Genesee Valley	L. Ontario—Allegheny R.	1840 (?)
Welland (Ship) Canal	Western L. Ontario—eastern L. Erie	1824
Trent Canal (Trent Waterway)	L. Ontario—L. Huron	1918
Ogoki Diversion	L. Superior—Hudson Bay	1943
Long Lac Diversion	L. Superior—Hudson Bay	1938

* Data mostly from: "A Description of the Canals and Railroads of the United States, Comprehending Notices of All the Works of Internal Improvement throughout the Several States" by H. S. Tanner, T. R. Tanner and J. Disturnell, New York, 1840, vii + 273 pp.

stragglers also enter through canals. Apparent failures were plants of the coho salmon, *Oncorhynchus kisutch* Walbaum, and the European whitefish, *Coregonus lavaretus maraena* (Bloch). The northern bullhead minnow presumably was introduced by bait fishermen. Within the Great Lakes basin, as elsewhere, ranges of several native minnows have been spread as a result of dumping surplus bait.

The total number of kinds of fishes, native and introduced, which are now known to live in the Great Lakes and their tributaries, or to have occurred there in historic times, is 234. Growth in our knowledge and understanding of this fish fauna is indicated by the changes in the number of families, genera, species and total forms including subspecies, as listed in successive treatises (Table V).

TABLE V

CHANGE IN NUMERICAL STATUS WITH TIME, FOR GREAT LAKES FISH GROUPS

Groups	*Number Listed by Various Authorities*				
	Evermann (1902)	Hubbs (1926)	Hubbs and Lagler (1939)	Hubbs and Lagler (1947)	Present Work
Families	(not cited)	28	28	29	29
Genera	74	96	93*	94	75†
Species	136§	154	167	173	173
Total forms, including subspecies	152	166	223	223	234

* Reduction due to a consolidation of the genera of sunfishes.

† This further reduction in number of genera recognized is due to a general though not complete acceptance of a current trend to lump genera—a trend that we think is at least in part desirable, though it may in part be leading toward an undesirable extreme, and is certainly introducing great change in nomenclature.

§ Including several species not now recognized as valid.

The families of Great Lakes fishes represent many different major groups (Table VI). In the American literature of ichthyology for the past three-quarters of a century, these families have been arranged into about seventeen orders, as in the system recognized by Jordan, 1929. The work of Berg, 1940, consolidated the same families into thirteen orders with systematized names. Firm in our convictions that the student of ichthyology needs to form concepts of the major natural groups (regardless of the names ascribed to them by different authors) we have incorporated many group names (largely ordinal ones of Jordan) in the key to the families that follows. The groupings are summarized and cross-referenced to the proposals of Berg, 1940, in Table VI.

TABLE VI
CLASSIFICATION OF FAMILIES OF GREAT LAKES FISHES

Family	*Orders (and Higher Groups)* *Jordan, 1923*	*Berg, 1940*
	(Class MARSIPOBRANCHII)	(Class PETROMYZONES)
PETROMYZONTIDAE Lamprey Family	Hyperoartia	Petromyzoniformes
	(Class PISCES) (Subclass ACTINOPTERI)	(Class TELEOSTOMI) (Subclass ACTINOPTERYGII)
ACIPENSERIDAE Sturgeon Family	Glaniostomi	Acipenseriformes
POLYODONTIDAE Paddlefish Family	Selachostomi	
AMIIDAE Bowfin Family	Halecomorphi	Amiiformes
LEPISOSTEIDAE Gar Family	Holostei	Lepisosteiformes
CLUPEIDAE Herring Family		
OSMERIDAE Smelt Family		
THYMALLIDAE Grayling Family		
SALMONIDAE Salmon Family	Isospondyli	
COREGONIDAE Whitefish Family		Clupeiformes
HIODONTIDAE Mooneye Family		
UMBRIDAE Mudminnow Family		
ESOCIDAE Pike Family	Haplomi	
CATOSTOMIDAE Sucker Family		
CYPRINIDAE Minnow Family	Eventognathi	Cypriniformes
ICTALURIDAE North American Freshwater Catfish Family	Nematognathi	
ANGUILLIDAE Freshwater Eel Family	Apodes	Anguilliformes
CYPRINODONTIDAE Killifish Family		
POECILIIDAE Livebearer Family	Cyprinodontes	Cyprinodontiformes
GADIDAE Cod Family	Anacanthini	Gadiformes
PERCOPSIDAE Troutperch Family	Salmopercae	Percopsiformes
APHREDODERIDAE Pirateperch Family	Xenarchi	
ATHERINIDAE Silverside Family	. . . .	Mugiliformes
SERRANIDAE Bass Family		
CENTRARCHIDAE Sunfish Family	Percomorphi	
PERCIDAE Perch Family		Perciformes
SCIAENIDAE Drum Family		
COTTIDAE Sculpin Family	Cataphracti	
GASTEROSTEIDAE Stickleback Family	Thoracostei	Gasterosteiformes

FIELD STUDY AND THE COLLECTION OF FISHES

"In the collection of fishes, three things are vitally necessary—a keen eye, some skill in adapting means to ends, and some willingness to take pains in the preservation of material." (David Starr Jordan, 1903.)

In the early days of ichthyology in North America, collectors were content to preserve only a few specimens of each kind of fish which they caught. The modern tendency is to obtain large series of all fishes encountered, common and uncommon forms alike, in order that materials for variational, growth and other studies may be available. This means that adequate containers, preservative and other materials must be carried into the field for scientific collecting. For hobby collecting, however, only a small amount of simple equipment is needed.

Containers may be almost anything—glass, wood or noncorrosive metal (preferred for periods of long storage). Tin cans, paint buckets and large or small oil cans are advantageous containers for shipping. Receptacles for actual use during collection should have an opening large enough to receive strongly compressed forms, such as sunfish, without injury to the fins. "Ball Eclipse" (extra-wide-mouth) jars are particularly suitable and a large container with a ten-inch opening will often be useful.

In most regions, the collector relies upon his own ability and techniques to seine, net or otherwise capture fish. In some places, however, he may be chiefly dependent on commercial fishermen.

In his own right, the collector may use any means to capture fish for which he can obtain sanction from local authorities. He may bait- or fly-fish, spear or gig, seine, net, trap, poison, electrocute, stun with explosives, use a set-line, shoot or even collect fish by shooting fish-eating animals. Effort should be made, however, to avoid arousing antipathy of conservation officials, sportsmen and land owners. It is well to notify local conservation officials before starting to collect in any region. Besides, they will often be able to make helpful suggestions.

For ordinary study collecting, "Common Sense" minnow seines in sizes of six-, ten-, twelve- and fifteen-foot lengths are quite practicable. More serviceable nets are tied ones, and of these a bag type, twenty-five feet long, is most widely employed. Except for special work, one-quarter-inch square mesh and seine depths of four to six feet are acceptable. The so-called "experimental" gill net has some utility in general collecting. This net is 125 feet long and five feet deep. It is composed of five twenty-five-foot sections, one each of the following square mesh sizes: ¾ in.; 1 in.; 1¼ in.; 1½ in.; 2 in.

In some situations, where seining is impracticable because of the nature of the bottom and where permision can be obtained, poisons may be used. Great care should be exercised, however, to avoid harm to the fauna or adverse public reaction. Powdered derris root with rotenone as its chief active ingredient is an effective poison in a concentration of 0.5 part per million parts of water by weight. Rotenone is advantageous because it does not harm humans, livestock or other animals in the concentrations in which it is likely to be used. Field units for the electrocution of fish have been devised although care must be taken to recover forms which, lacking an air bladder, do not rise to the surface when electrocuted.

The greatest concern of the collector should ever be that he neglect no habitat in the aquatic community under his survey—that he assume fish to be everywhere in the water, on all bottoms and at all depths, until he has proven otherwise to his own satisfaction. An experienced collector is wont to say, "Where there's water there's fish and where there's fish we catch 'em." Riffles, pools, weed beds, undercut banks, shoals and deeps should each be attacked with the proper techniques and gear for getting fish from them. Operations may be conducted profitably at all times of day and night and in all seasons. Night seining is generally more effective than day seining, particularly for deep-water species and large adults of other kinds. The young (ammocetes) of lampreys must be dredged from mud in stream banks. Parasitic lampreys can often be obtained from their hosts. In certain regions caves and wells should not be neglected.

Detailed notes made at the time of each collection greatly enhance the value of a set of fish specimens. Such notes should describe the ecological relationships of the fishes encountered. The headings of a sample sheet for field notes are shown in Fig. 2. This form when printed on the upper third of a sheet of medium weight bond paper about 5½ by 7½ inches will serve as a guide in recording minimum standard information and will leave space

Coll. No. ....

State or Country: .... **County** ....

Locality: ....

....

Water: ....

Vegetation: ....

....

Bottom: .... **Temp.:** .... **Air:** ....

Shore: .... **Current:** ....

Distance from shore or stream width: .... **Tide:** ....

Depth of capture: .... **Depth of water:** ....

Method of capture: ....

Collected by: .... **Date:** ....

Orig. preserv.: .... **Time:** ....

Fig. 2. Headings for field sheet used in recording fish collection notes.

below for additional ecological observations and for special notes on individual species. Further entries can be made to record habits, to list discarded specimens and to describe bright colors that fade after preservation. All writing should be done with waterproof carbon ink (such as "Higgins Engrossing") that will not be destroyed by wetting, or with a typewriter equipped with a good permanent black ribbon. Copies of field notes should ultimately be filed where the collections are housed for study.

Suggestions regarding the reporting of data follow under the appropriate headings from the field sheet.

Coll. No. The assignment of a serial number to each collection facilitates the identification of the specimens or container from a particular place and time. Precede the number by the collector's initial (or other indication of a series of collections) and the year to avoid duplication. In the marking of containers, it is well to write the collection number, locality data, date and collector on a label that is kept inside the container and at least a numbered tag on the outside. Several grades of heavy-weight bond paper make good label stock. One kind that has proved very satisfactory is "L. L. Brown's Resistall Linen Ledger, Substance 36."

State or Country and County. These items are given first because distributional records are often filed or summarized according to the political entities in which the collections were made. Local workers are apt to neglect this information, which others may need.

Locality. Include here: name of body of water, if it has one; the stream system; exact geographical location on the water area in which the collection was made; and reference in miles and direction to a nearby town, using a larger one in preference to a smaller one. To execute these requirements one must have on hand an accurate map of sufficiently large scale—one-half inch to the mile is a suitable minimum. A good general rule is to state the locality in general terms so that it may be located readily with ordinary maps and also in sufficient detail so that if occasion arises one could return to the approximate spot to repeat the collection.

Water. Write here information on clearness of the water. Commonly used categories are clear, slightly turbid and turbid. Include also notes on color. Color refers to stain in the water itself, not to apparent color imparted to it by suspended matter. Ordinary conditions encountered are colorless, light brown and brown. The latter colors are mostly associated with bog waters. Other colors suggest pollution. Hardness and other chemical features should be noted, if known. In regular surveys turbidity is often expressed by the depth where a Secchi disc disappears from view. In other collecting it is well to state the "bottom visibility"—the depth where the bottom is barely seen.

Vegetation. List kinds of water plants with an indication of their degree of relative abundance (as sparse, medium or dense) and their distribution. Vegetation is important for it is the principal habitat of certain fishes and affords cover and feeding grounds for these and others.

Bottom. Describe nature of bottom materials and give relative abundance and location where this is pertinent. Common bottom types in this region are: silt, sand, gravel, rubble, boulders, bedrock, clay, marl, pulpy peat, fibrous peat and detritus (trash, sticks, etc.). Certain species will be seen to show distinct preferences for a particular substratum.

Temperature. A good pocket thermometer, about the size of a fountain pen, in a metal case is a minimum requirement for ordinary fish collecting. If practicable, it should be calibrated against a standardized instrument. Temperature of air and water at the time of collections help identify thermal habitats such as cold and warm or hot springs. Temperatures, of course, are very important in determining where certain fishes will live.

Shore. Reference should be made to the nature of immediate shore as related to the aquatic habitat, then to surrounding country.

Current. Nature of the current may be expressed as relative rate of flow; stagnant, sluggish, slow, moderate, rapid and torrential are terms frequently used. Commonly "pools and riffles" will be the entry.

Distance from shore or stream width. In setting traps or gill nets, or in open-water seining, distance from shore is often a significant descriptive measure. In streams, least, greatest and average widths constitute the minimum notations.

Tide. In the oceans, as well as in the Great Lakes proper, tidal conditions affect the distribution of fish and, thereby, the kinds obtainable by netting a particular ground. Stage of the tide may be given as rising, high, ebbing or low. In low-tide collecting, the level attained should be entered.

Depth of capture and depth of water. Depth of water is always a needed entry, but depth of capture is entered only when specially known. Thus a fish may be captured at the surface or on the bottom, where the depth is 50 feet.

Method of capture. The method of collecting determines, in part, the species one will get from a particular habitat. To indicate the limitations of one's technique as well as to facilitate comparisons of one collection with another, a complete description of the gear (and sometimes of its operation) is necessary.

Collected by. Names of collectors or collecting party are given here and customarily one gives first the name of the person responsible for the notes.

Original preservative. The original preservative is recorded so that subsequent treatment of the specimens may proceed from a known basis. This is of particular significance if a specimen comes to be used for histological purposes requiring further chemical treatment.

Date. Always enter the month, preferably as a Roman numeral, before the day (as VI:25:1946).

Time. The time interval has reference to period actually spent in collecting. This record is useful in detecting and interpreting temporal changes in the fauna, particularly with reference to day and night. Furthermore, the length of time spent in making the collection may be a partial indication of its adequacy. Weather and solar or lunar stages may be noted, when pertinent.

Notes. In writing up the species list for each locality, future reference is facilitated if the fishes collected are recorded in family sequence. Under each, consideration should be given to the following points of information: (1) correct identity; (2) relative degree of abundance—such as rare, common or abundant; (3) life history stage—such as young, juvenile, adult, or, when known, young-of-the-year, yearling, two-year-old, etc.; (4) natural history notes—for example, stage of maturity (immature, mature, ripe, etc.), reproductive habits, food and feeding activities; (5) ecological notes—such as "in pools," "under overhanging banks," "on riffles," "at surface;" (6) detailed color descriptions, particularly of breeding fish.

PRESERVATION OF FISHES FOR STUDY

Not all fishes can be preserved whole in an embalming fluid. Large specimens may be skinned and preserved dry or in liquid. If dry, the skin should be thoroughly cleaned from flesh and fat and well salted. The brains should be removed, tooth bearing bones of pharyngeal region preserved and the skin completely desiccated under a slight weight to prevent undue shrinkage. If the specimen is to be "pickled," the skinning need not be as complete (the middle third of the carcass only need be removed) and the head and pharyngeal region as well as the tail may be left intact. Trophy specimens should be handled by a competent taxidermist.

The most widely used field and laboratory fixative for fish is made by mixing ten parts of water with one part of formalin (a saturated aqueous solution of formaldehyde gas, about forty per cent formaldehyde by weight). The strength may be increased or decreased under certain conditions. Extra large fish require a stronger solution (about eight parts of water to one of formalin); very small ones are preserved in more dilute formalin (about fifteen to one). The addition of household borax (one level teaspoonful to a full quart of preserving solution) has been found to retard shrinkage of specimens, the hardening of the soft parts and the softening of bony tissues. In place of borax, other buffering chemicals may be used. An excess produces an edema-like puffiness.

In general, specimens more than a few inches in length should have an incision made on the right side of the abdomen to facilitate penetration of the preservative. The right side is chosen for this and any other operations of a similar nature, such as the removal of scale samples, because the left side of the fish is ordinarily used for obtaining morphometric data and is commonly shown in photographs. The cut should be about half as long as the body cavity and should be made with a very sharp knife. The indiscriminate use of a hypodermic syringe for injecting formalin is discouraged. Not only do hypodermic injections often alter the body shape more than incising, but subsequent slitting is necessary if the formalin is to be adequately soaked from such specimens and if the internal organs need be examined. When fish heavier than three pounds are being prepared for preservation, it is desirable to make a deep incision into the muscle mass on each side of the vertebral column, operating from inside the body cavity.

For formalin fixation of the average specimen two days to a week is sufficient. For optimum permanent preservation, formaldehyde is removed from specimens as soon as possible after fixing by soaking them in water. Such material should be soaked for at least two days and the water changed at least once during this period. Transference is recommended to seventy per cent ethyl alcohol or, even better and much cheaper, to forty per cent isopropyl alcohol, for the removal of the remaining traces of formalin. One change of alcohol is needed before permanent storage in fresh alcohol of the strength stated above. Colors fade less rapidly when specimens are kept in darkness. Colors may be preserved for prolonged periods of time in special, usually more costly, preservatives. If formalin must be used as a permanent preservative, it should be used in dilute solution (one to fifteen) with borax or other buffer added as previously indicated.

If it is necessary to handle formalin specimens, immersal for a few minutes in a solution made up to the following specifications will remove the objectionable fumes: 4.5 gallons of tap water plus 1260 grams of $NaHSO_3$ and 840 grams of Na_2SO_3. Dissolve the salts in the tap water and add enough water to make five gallons of solution. This deodorizing bath will service many specimens before it needs to be replaced.

IDENTIFICATION OF FISHES

For fishermen and for amateur or professional ichthyologists the accurate identification of fishes and a knowledge of their distribution, habits and habitats are of great importance. We have here attempted to provide means by which all persons interested may: (1) learn characters of importance for the identification of fresh-water fishes and the means by which they may be accurately distinguished; (2) "key" the Great Lakes fishes to their correct names; (3) become acquainted with their known geographic ranges; and (4) find a summary of natural history information for each family and an ecological annotation for each species. The information that we have included on methods for collecting and preserving fishes will prove useful to the technical worker and to the hobbyist collector. The angler may also wish to save a fish for future identification since a microscope is often needed for examining small specimens.

Line drawings, color plates and photographs, augmented by the directions for making counts and measurements, provide a basis for learning salient characters of fishes and also for training one's self in the methods currently in use for their classification. The student should become thoroughly acquainted with this material before attempting to use the keys since the latter are based on systematic data obtained by the means described.

The keys given are for the most part artificial, in that the distinguishing characters used and the order in which the forms are treated do not usually indicate phylogeny. In the distributional list of species which follows the key for each family, however, an attempt has been made to align the units according to their relationship insofar as this is possible in a running list.

The keys have been made as simple as possible, without an undue sacrifice of the means for precise determinations. For certain families, however, it will be found expedient to consult the references indicated. Such works will be found to contain materials which will facilitate and verify identifications.

In "running" a fish through the keys, the following procedure should be used by the beginner, but may be modified by the advanced worker to suit his knowledge and the practices to which he is accustomed. Throughout the routine, careful checks should be made against the text figures and the section on methods to insure correct interpretation of the key characters.

(1) Determine family in "Key to the Families."

(2) Identify to lowest taxonomic unit listed in key to the family of which the fish is a member.

(3) Verify final determination by ascertaining:

(*a*) that the geographic range as given in the list includes the locality from which the specimen was taken;

(*b*) that specimen looks like the illustration;

(*c*) that the example at hand matches specimens previously identified by one qualified to do so;

(*d*) that it corresponds with the description of the species in Jordan and Evermann (1896-1900), in Forbes and Richardson (1920), in Hubbs (1926) or in some other standard work (see List of References).

The characters in the keys have been drawn from the Great Lakes representatives of the fishes in question. This work should therefore not be employed for the identification of fishes in any other area, except (with caution) for regions immediately adjacent to the Great Lakes and their tributary lakes and streams.

No provision is made in the keys for the identification of intergeneric or interspecific hybrids or of intergrades between subspecies. Hybrids and intergrades occur commonly in nature and may usually be recognized by the circumstance that their characters are intermediate between those of the parental forms.

The geographic range as we have given it for each species and subspecies has been compiled, with an attempt at precision, from a vast store of original as well as published data. Much of the zoogeographical information in this bulletin is therefore new. Many of the ranges are extended to regions and localities from which no previous records have been made known. Such new information is chiefly based on the extensive collections which are actively being accumulated and studied at the Museum of Zoology, University of Michigan. All of the many source references used from the literature are not cited but the most important ones are included in the List of References.

The notes as to usual habitat are oversimplified for the sake of brevity. Important additional ecological information and observations on habits, economic value and significance for sport are given in other publications, many of which are listed subsequently in this bulletin.

ANATOMICAL FEATURES, AND TERMS AND METHODS OF COUNTING AND MEASURING

(Figs. 3–7)

There has been much variation in the methods of counting scales, fin rays and other serial parts and in the methods of measuring body lengths and other dimensions. Workers have altered procedures to suit investigational needs, the nature of the material or personal whims, but have seldom stated how the methods employed by them differ from more standard ones. The methods here given are widely used but do not invariably conform with past practices or with those in use by all other workers. Notes are provided which permit coördination. Several of the procedures described are not employed in the keys that appear later in this book. They are included, however, because they afford means for describing and comparing differences among fishes. Also given are salient features of external and internal anatomy, a knowledge of which is necessary for making identifications.

Methods of Counting Fin Rays

Number of Fin Rays (Fig. 3, below)—In specifying the number of fin rays, "dorsal rays," "anal rays," etc. are either written out of the following nomenclature and abbreviations are used:

Dorsal fin rays—D	Pectoral fin rays—P_1
Anal fin rays—A	Pelvic (ventral) fin rays—P_2
Caudal fin rays—C	

For paired fins the counts are made on the left side of the body unless otherwise specified.

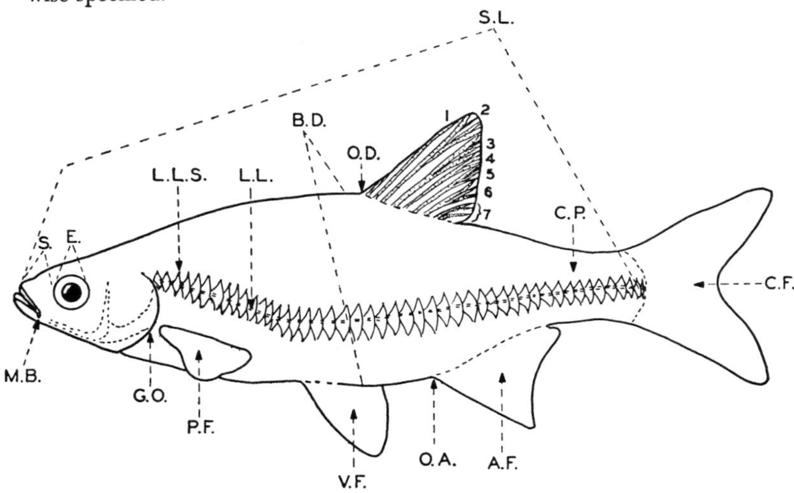

Fig. 3. Topography of a hypothetical soft-rayed fish (from Hubbs and Cooper, 1936). A. F., anal fin; B. D., body depth; C. F., caudal fin; C. P., caudal peduncle; E., length of eye; G. O., gill-opening; L. L., lateral line; L. L. S., lateral line scales; M. B., maxillary barbel (in terminal position); O. A., origin of anal fin; O. D., origin of dorsal fin; P. F., pectoral fin; S. length of snout; S. L., standard length; V. F., pelvic fin; 1 to 7, dorsal fin rays; (note that two very small rays at the front end of this fin are not counted and that the last fin ray is divided through its base).

Spines. All true spines (Fig. 4, D_1, below) are designated by Roman numerals no matter how rudimentary or how flexible (Figs. 14, p. 28 and 50, p. 118) they may be. It is desirable to treat numerically as spines the morphologically hardened soft-rays, whether these be simple rays as in the carp (*Cyprinus carpio*—Fig. 11) or the consolidated product of branching as in catfishes (Ictaluridae). True spines are median (unpaired) structures, without segmentation.

Soft-rays, designated by Arabic numerals, are usually though not always branched and flexible, and are bilaterally paired and segmented (Figs. 3, p. 19 and 37, p. 71).

Spines and soft-rays in one fin. In a fin containing both spines and soft-rays (Figs. 16 and 17, p. 28) the count for the spines is separated from the soft-ray count by a dash, if the two sections of the fin are separated. If the two sections are conjoined, a comma is used to separate the counts.

Ray. The term "ray" designates spines as well as soft-rays (and the latter term may be hyphenated to give it technical significance).

Principal and branched rays (Fig. 3, p. 19). In certain fishes, particularly the Cyprinidae and Catostomidae, the count is of principal rays, to accord with general practice and because the rudimentary rays are difficult to ascertain or are variable. Almost without exception in these families, the principal rays include the branched rays plus one unbranched ray, since only one unbranched ray reaches to near the tip of the fin. If it is desired to give the count as of the number of branched rays, the term "branched rays" is employed, but the principal ray count is recommended for use.

Rudimentary rays. In groups of fishes such as Ictaluridae, Esocidae and Salmonidae, in which the "rudimentary" rays grade into the developed ones, both in length and degree of branching, the total count is given. And when the principal rays are enumerated, this is done by adding one to the count of the number of branched rays, including as branched any ray that is at all forked. The maximum total count is given for all fins in which few or no rays are branched.

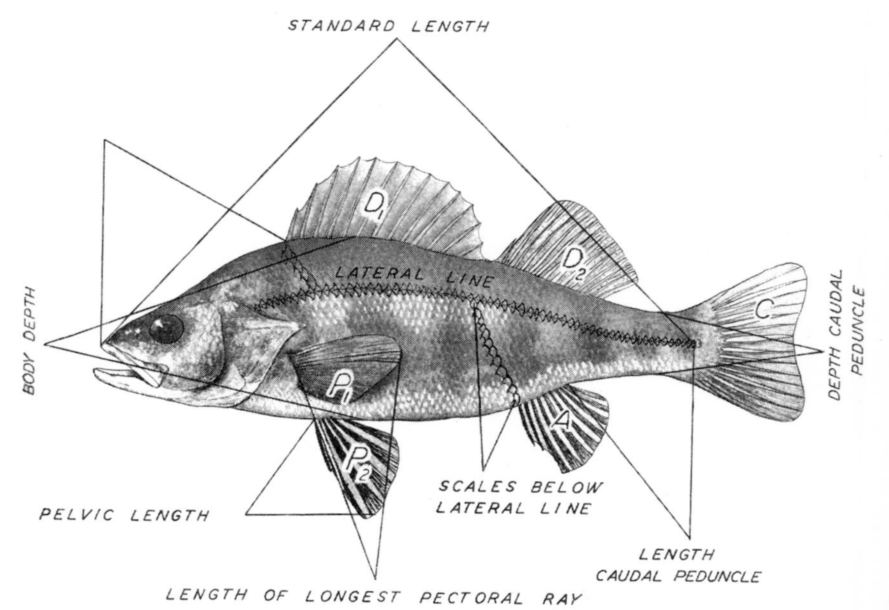

Fig. 4. Topography of a spiny-rayed fish (yellow perch, *Perca flavescens*), showing how certain measurements are made and locating structures and regions used in identification.

Last ray of dorsal and anal fins (Fig. 3, p. 19). In the dorsal and anal fins the last ray, for the purpose of the count, is defined as consisting of two ray elements that are separated (even though serially approximated) at the very base of the fin. In other words, the last two bases are counted as one ray. (This has been the general custom for counting fins in which the rays are well branched; there has been little consistency in this regard in the counts for fishes having the rays unbranched; no single, simple definition other than that given above would seem sufficient to designate the last dorsal and anal fin element that is to be counted as one ray.) In some special studies it has proved advantageous to compare the numbers of branched and unbranched rays.

Caudal rays (Fig. 10, p. 28). Ordinarily the caudal fin count as given is that of the principal rays. In fishes having branched caudal rays, the number of principal rays is defined as the number of branched rays plus two (for this is the obvious count).

Rays in paired fins. In the paired fins, all rays are counted, including the smallest one at the lower or inner end of the fin base. Very often good magnification is needed in this count. Frequently a small ray (counted in the pectoral but not in the pelvic fin) precedes the first well developed ray and may be bound very closely to it so as to require dissection to be seen.

In certain fishes with reduced pelvics, such as the Cottidae, the spine may be represented by a mere bony splint bound into the investing membrane of the first soft-ray (Fig. 50, p. 118), which can be recognized as such under the microscope by the articulations and by the bilateral structure.

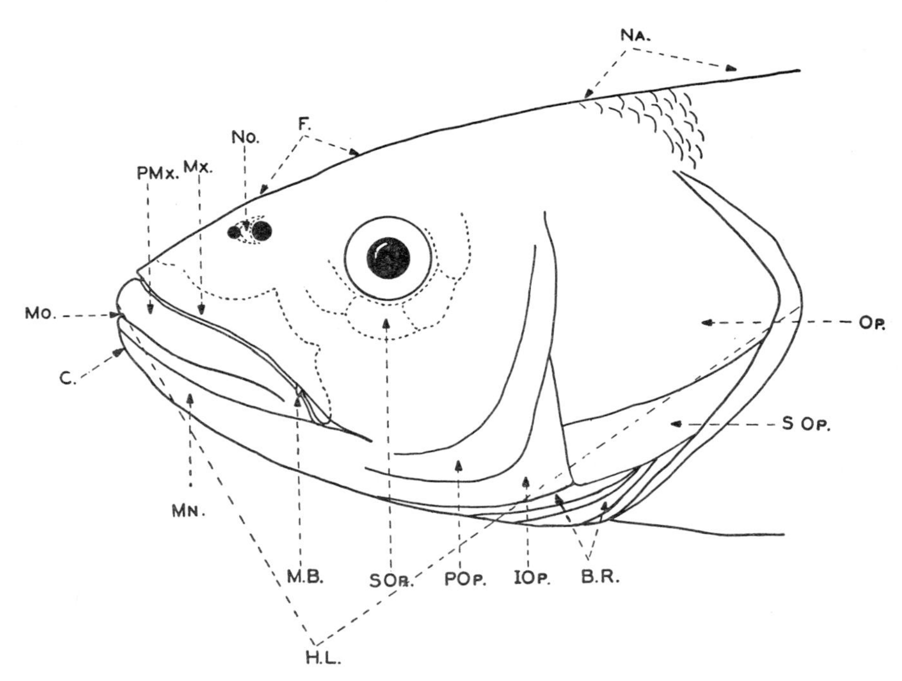

Fig. 5. Head of a soft-rayed fish (northern creek chub, *Semotilus a. atromaculatus*), showing structures and regions used in identification (after Hubbs and Cooper, 1936). B. R., branchiostegal rays; C., chin; F., forehead; H. L., length of head; IOp., interopercle; M. B., maxillary barbel (set forward on maxillary); Mn., mandible; Mo., mouth; Mx., maxillary; Na., nape; No., nostrils; Op., opercle; PMx., premaxillary; POp., preopercle; SOp., subopercle; SOr., suborbitals.

In general, the maximum possible scale count is stated, including small interpolated scales in the lateral line and the scales of reduced size near the origins of the vertical fins, but not including the scales on the fin bases or on the basal sheaths.

Lateral line scale count (Figs. 3, p. 19 and 4, p. 20) represents the number of pores in the lateral line or the number of scales along the line in the position which would normally be occupied by a typical lateral line. Count terminates at the structural caudal base or end of the hypural plate, as determined without dissection by moving the caudal fin from side to side. If the crease between the caudal fin joint and the body underlies a scale, the question of inclusion or exclusion is determined by the test of whether the crease appears to lie behind or in front of the middle of the exposed field of that scale. The scales wholly on the caudal fin base are not included in the count, even when they are well developed and pored. Sometimes referred to as "scales in lateral line" or as "scale rows along side of body."

The most anterior scale enumerated is that one which is in contact with the shoulder girdle but is followed by one which is definitely separated by another scale from the shoulder girdle. That is, in counting forward, the last scale counted is the first one to touch the shoulder girdle.

Scales above lateral line (Figs. 4, p. 20 and 6, p. 22). Unless otherwise indicated the count of scale rows above the lateral line is taken from the origin of the dorsal fin (or from the origin of the first dorsal fin if there is more than one), including the small scales, and counting downward and backward following the natural scale row to, but not including, the lateral line scale.

Scales below lateral line (Fig. 4, p. 20). The count of scale rows below the lateral line is taken similarly to that for rows above the lateral line. The

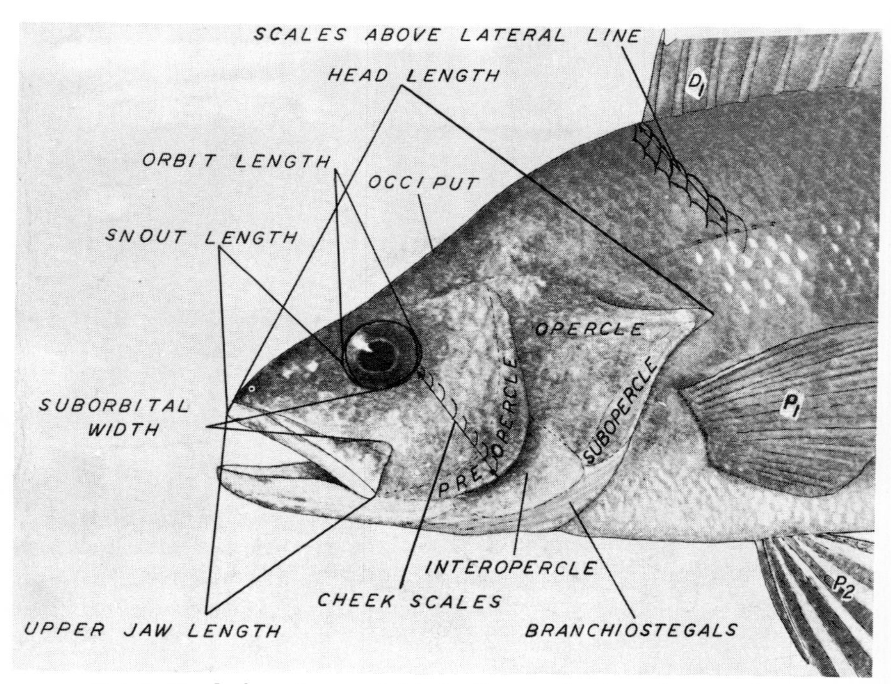

Fig. 6. Head of a spiny-rayed fish (yellow perch, *Perca flavescens*), showing topographical features and how certain measurements are made in identification.

count is made upward and forward from the origin of the anal fin. In this count, as in the one above the line, the small scales are included. If in continuing upward and forward the series can with equal propriety be regarded as jogging backward or forward the backward shift is accepted. The scale nearest the fin is counted as one-half only when this is very definitely an evident characteristic.

Scales before the dorsal fin. The number of scales before the dorsal fin is determined by counting all scales, the exposed surfaces of which wholly or partly intercept the straight midline running from the occiput to the origin of the dorsal fin. Ordinarily this count is made in fishes in which the transverse occipital line very sharply separates the scaly nape from the scaleless head. The "number of scale rows before the dorsal" (commonly fewer than the number of predorsal scales) is made to one side of the midline.

Check scales. This count represents the number of scale rows crossing an imaginary line from the eye to the preopercular angle. (Fig. 6, p. 22)

Circumference scale count (particularly valuable in the Cyprinidae) represents the number of scale rows crossing a line around the body immediately in advance of the dorsal fin.

Caudal peduncle scale count is taken similarly to the circumference scale count but is made around the part of the peduncle where the count is lowest.

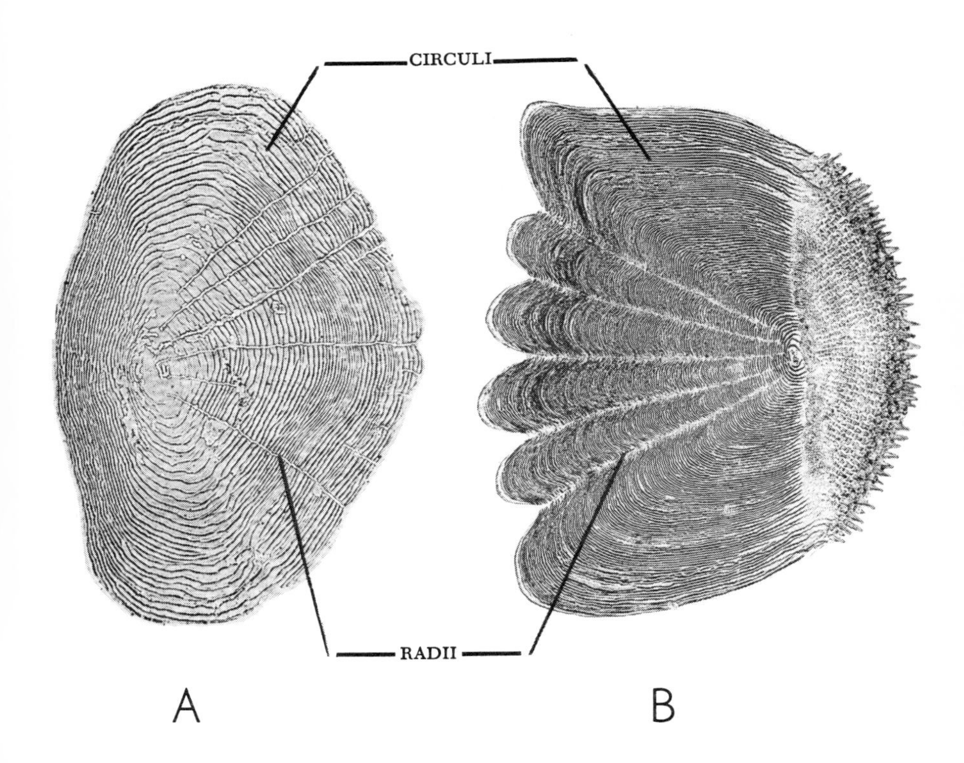

Fig. 7. Two common types of fish scales, with embedded portions to the left, and the posterior, exposed margin to the right. A. Cycloid scale (northern mimic shiner, *Notropis v. volucellus*). B. Ctenoid scale (yellow perch, *Perca flavescens*).

Other Counts

Branchiostegals or branchiostegal rays (Figs. 5, p. 21, 6, p. 22 and 40, p. 94). It is often desirable to separate by a plus sign (+) those branchiostegals which lie on the outer side of the hyoid arch from those that are inserted more anteriorly and more ventrally on the inner face of the arch. Care should be taken to include the most anterior branchiostegals which are apt to be very short, slender and concealed. The method of making this count has been described by Hubbs (1920).

Pharyngeal tooth counts (Fig. 36, p. 69). In minnows counts are made on the "throat-teeth" borne on two bones which are modified fifth gill arches and must be temporarily removed (with great care) and cleaned so that the count may be made. Each of these bones bears one or two rows of teeth (three rows in the introduced carp). The teeth in each row are counted and given in a formula in order from left to right; thus the formula 2, 5–4, 2 indicates that the pharyngeal bone of the left side has two teeth in the outer row and five in the inner, whereas the right bone has four teeth in the inner row and two in the outer. The formula 4–4 discloses that the fish has no teeth developed in the outer row. Pharyngeals of suckers are shown in Figs. 15, p. 28 and 32, p. 60.

Gill-rakers (Figs. 29, p. 51 and 49, p. 112). Unless otherwise stated, the count is that of the first arch. It is sometimes of value, however, to count the rakers on the other arches as well. A single gill-raker count indicates the number on the entire first arch, but has often been used for those on the lower limb only. If the numbers on the upper limb and lower limb are taken separately, the two figures are separated by a plus sign. If the count is taken along the lower limb only, the fact is stated. If a gill-raker straddles the angle of the arch, it is included in the count of the lower limb. All rudimentary rakers are to be included in the count (unless it is stated that the rudiments are excluded).

Pyloric caeca. In counting pyloric caeca all tips are enumerated unless the condition of branching is specifically described.

Vertebral counts. The typical hypural plate (Fig. 10, p. 28) of most teleosts is counted as a single vertebra. However, definite sutures along the vertebral axis are regarded as separating vertebrae, even though the suture or sutures lie within the hypural complex (this reservation applies particularly to the Salmonidae). In heterocercal (Fig. 8, p. 28) and abbreviate-heterocercal (Fig. 9, p. 28) tails, all elements are counted that are separated by definite sutures. Precaudal and caudal vertebrae are commonly distinguished. The first caudal vertebra is the first vertebra bearing a definite hemal spine. The last one to several precaudal (trunk) vertebrae may have complete hemal arches.

Methods of Measurements

Smoothly working dividers or dial-reading caliper should be used for measurements. Dividers should have one point flat at a right angle to the plane of operation and the other kept at a needle point. A steel ruler of good quality is recommended for precise readings. Great caution should be exercised in the way of accuracy. Measuring boards as commonly used in fishery investigations are hardly suitable for routine systematic work.

Unless otherwise stated, all measurements are taken in a straight line, from point to point rather than around the curve or as a projection. When the body or any part being measured has been curled, bloated, or otherwise distorted on death or in preservation, or when the head has been fixed in abnormal position, thrown upward and backward (in opisthotonus), with the gill-covers dilated, the part being measured is gently forced into as nearly the normal appearance as possible before being measured.

In descriptions it is customary to express the size of each part as a proportion of the standard length or of the head length, or occasionally of some other base. For routine descriptions the smaller part is conventionally divided into the larger, as head (length) 4.2 in standard length, or eye (length) 3.5 in head (length). These values are usually obtained by stepping the length of the part into the base length over the curve of the latter, and this is our recommendation. Some, however, make the division arithmetically; when that is done, the practice should be so stated. In variation studies and for precise descriptions, the size of the parts is expressed in hundredths, or, better, as thousandths of the standard length. The divisions are most readily performed on a calculating machine.

Total length is the greatest dimension between the most anteriorly projecting part of the head and the farthest tip of the caudal fin when the caudal rays are squeezed together. The measurement is a straight line and is not taken over the curve of the body.

Standard length. In fishery work, as a result of the use of the measuring board, the standard length is taken as the distance from the most anterior part of the head (whether the lower jaw or the upper jaw projects) backward to the end of the vertebral column (structural base of the caudal rays).

In systematic work the standard length (Figs. 3, p. 19, and 4, p. 20) is properly the distance from the most anterior part of the snout or upper lip to the caudal base (although this has not been universal practice). Since the measurement is a straight line it is not taken over the curve of the body.

Body depth (Figs. 3, p. 19 and 4, p. 20) is the greatest dimension, exclusive of the fleshy or scaly structures which pertain to the fin bases.

Depth of caudal peduncle (Fig. 4, p. 20) is the least depth of that part.

Length of caudal peduncle (Fig. 4, p. 20) is the oblique distance between the end of the anal base and the hidden base of the middle caudal ray.

Predorsal length is the distance from the tip of the snout or upper lip to the structural base of the first dorsal ray.

Length of dorsal or of anal base is the greatest overall basal length, extending from the structural base of the first ray to the point where the membrane behind the last ray contacts the body.

Height of dorsal or of anal fin is taken from the origin of the fin to the tip of the anterior lobe.

Length of depressed dorsal or anal is the distance from the base of the first ray to the farthermost point when the fin is flattened down.

Length of longest dorsal or anal ray is measured from the structural base of the longest ray to its tip.

Length of pectoral or of pelvic fin (Figs. 4, p. 20). The length of the paired fins is the distance from the extreme base of the uppermost, outermost or anteriormost ray to the farthest tip of the fin, filaments, if any, included. For the pectoral fin this measurement is used when the fin is asymmetrical.

Length of longest pectoral ray (when this ray is at or near the middle of the fin) is measured from the middle of the base of the fin (Fig. 4, p. 20).

Spine and soft-ray lengths. When a spine is being measured, especial care is required to make sure that one tip of the dividers is inserted at the very base of the spine. Soft-rays are measured to their most extreme tip, but spines are measured only to the tip of the spine proper, not including filaments or soft-rayed extensions (as on the false pectoral spine of catfishes).

Head length is the distance from the most anterior point on the snout or upper lip to the most distant point of the opercular membrane (Figs. 5, p. 21 and 6, p. 22). Many authors, however, have excluded the membrane from the measurement.

Depth of head is measured from the midline at the occiput *vertically* downward to the ventral contour of the head or breast. If the cross-line of the isthmus is distinctly removed from this vertical, then a measurement "occiput to isthmus" may be taken.

Head width is the greatest dimension when the opercles, if dilated, are forced into a reasonably normal position.

Snout length (Figs. 3, p. 19 and 6, p. 22) is taken from the most anterior point on the snout or upper lip to the front margin of the orbit.

Postorbital length of head is the greatest distance between the orbit and the membranous opercular margin.

Suborbital width (Fig. 5, p. 21) is the least measurement from the orbit to the suborbital or preorbital margin.

Height of cheek is the least distance from the orbit downward to the lower edge of the anterior arm of the preopercle.

Length of cheek is the distance from the most posterior point of the preorbital (lachrymal) horizontally backward to the caudal margin of the preopercle, including spines if present approximately along this horizontal.

Orbit to angle of preopercle. The distance from the orbit to the angle of the preopercle is taken to include any spine at the angle.

Interorbital widths. In determining the *least fleshy width* of the interorbital, the dividers are not squeezed at all, but in measuring the *least bony width,* the points are pressed tightly against the bone so as to eliminate so far as practicable the thickness of the flesh overlying the bony rims.

Length of orbit (Fig. 6, p. 22) is the greatest distance between the free orbital rims, and is often oblique.

Length of eye (Fig. 3, p. 19), as contrasted with length of orbit, is the greatest distance across the cornea, that is, between the margins of the cartilaginous eye-ball. The location of the margin can be determined rather accurately by close examination, and by touching the eye surface with the points of the dividers, thereby causing the margins to become more visible, since the cornea is thinner and softer than the eye-ball.

Length of upper jaw (Fig. 6, p. 22) is the term that now replaces "length of maxillary," which is not truly descriptive since the measurement is taken from the anteriormost point of the premaxillary to the posteriormost point of the maxillary.

Length of mandible. In measuring the mandible, one tip of the dividers is inserted in the posterior mandibular joint, so as to give the maximum possible dimension.

Width of gape is the greatest transverse distance across the opening of the mouth.

Fish Names

Throughout this book, great care has been taken to give the user the most accurate, acceptable and up-to-date names for the many fishes treated. Two sets of names are involved, technical (scientific) and lay (common). Unhappily, there is only moderate stability in both; as new information is gained and as new concepts form, changes inevitably come about. Fish students, acquainted with earlier editions of this bulletin, will be aware of many of the changes that we have incorporated and will also know of many others that are pending. Although distressing to one who must continually re-learn names, or to one who must look under several names, when bibliographing a fish, change is the very essence of progress, and also the very clear manifestation that the science of ichthyology is a live one. Furthermore, the fact that there is change may be taken as encouragement by students. Even in such a circumscribed and much-investigated fauna as this, far from all of the taxonomic problems have been solved. Future works on Great Lakes fishes will always have in them some differences in naming (and classification) from those presented herein.

Each scientific name of a fish is composed of two parts and is latin or in latinized form. One of these names, the first, is the genus or generic name. The other, the second, is the species name. Thus, each of the families of fishes in the Great Lakes has one or more genera in it. Similarly, each genus has one, but may have more than one species in it. And species are divisible into groupings of close relatives called subspecies. When subspecies names are used, the scientific name is of three parts, or tri-nomial. An example follows for the northern smallmouth bass:

Genus	Species	Subspecies
Micropterus	*dolomieui*	*dolomieui*

When a scientific name of a fish is written, as in the keys and statements of ranges in this book, it is followed by the name of the person who first properly ascribed the scientific species (or subspecies, for tri-nomials) name to that kind of fish. Subsequent workers may have technical ground for moving this species to a genus other than that in which it was originally described. When this is done, the name of the first describer accompanies the species designation to the new generic location but is placed in parentheses to call attention to the fact that a shift has been made.

Alterations in the usage of scientific names, including shifts of species from one genus to another, are sometimes the result of painstaking study—research in greater detail and thoroughness than has been done or has been possible in previous time. Such emendations are welcome and tend to be durable. At other times, changes are made on a subjective basis and are less likely to be true and, therefore, acceptable. Our tendency has been to be conservative regarding change in this edition, with the full realization that certain proposals which we have deigned to follow may in the future be strengthened to the point of acceptability through additional study.

Alterations in scientific names and their assignment and usage follows established rules of the International Zoological Congress. Unfortunately, there is no such set procedure regarding common names. It is not surprising, therefore, that the beginning student and informal fisherman may be confused greatly by the profusion of common names that are applied to one and the same fish at different points in the Great Lakes basin and elsewhere. A classical example is that of the yellow walleye (*Stizostedion v. vitreum*) which is reputedly known by more than 80 common appellations throughout its range in chosen parts of the United States and Canada. Considerable effort is going into standardization of common names for use in print (even though highly localized names will continue their provincial existence). Most active and cooperating in this regard are the Outdoor Writers of America, the American Society of Ichthyologists and Herpetologists, and the American Fisheries Society. Most of the changes in common names in this edition, compared to the previous ones, result in our desire to conform with the printed results of the efforts of these groups.

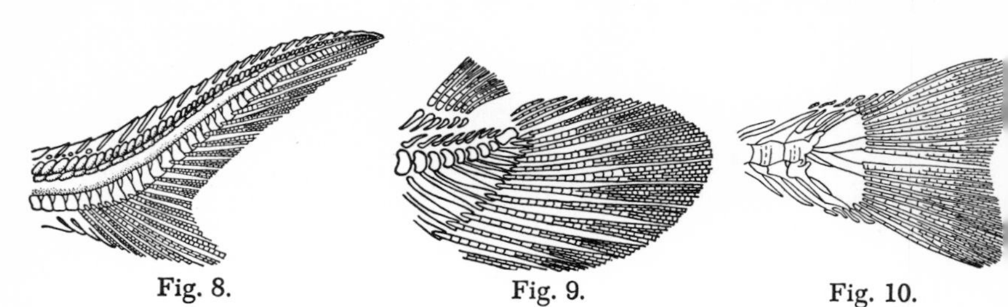

Fig. 8. Typically heterocercal tail (lake sturgeon, *Acipenser fulvescens*).
Fig. 9. Abbreviate-heterocercal tail (bowfin, *Amia calva*).
Fig. 10. Homocercal tail (northern largemouth bass, *Micropterus s. salmoides*).

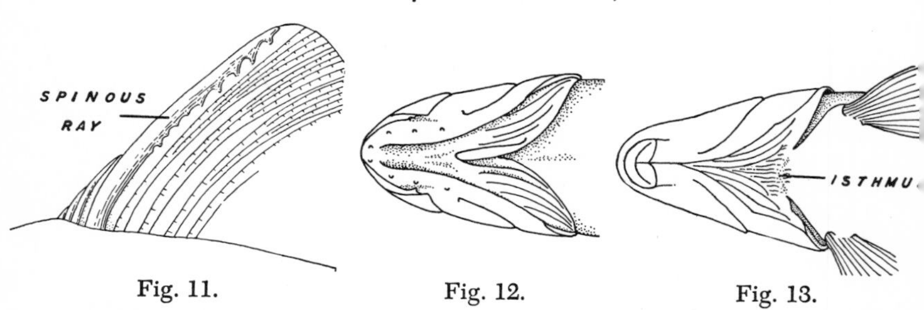

Fig. 11. First principal ray (strong, spinous and barbed) in anterior portion of dorsal fin of carp, *Cyprinus carpio*.
Fig. 12. Gill-membranes free from isthmus (northern rock bass, *Ambloplites r. rupestris*).
Fig. 13. Gill-membranes united to isthmus (common white sucker, *Catostomus c. commersonnii*).

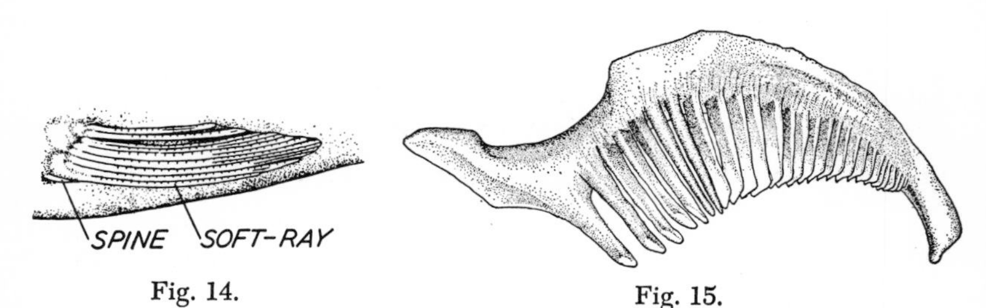

Fig. 14. Ventral view of right pelvic fin of troutperch, *Percopsis omiscomaycus*, with skin removed from outer edge to show small spine which might easily escape notice.
Fig. 15. Pharyngeal arch of golden redhorse, *Moxostoma erythrurum*, showing comb-like teeth.

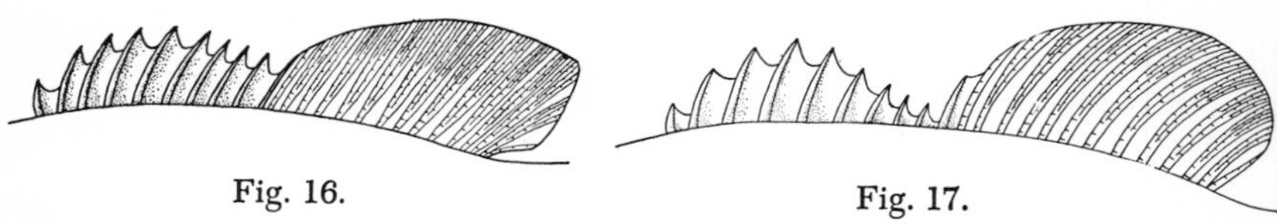

Fig. 16. Dorsal fin single (northern smallmouth bass, *Micropterus d. dolomieui*).
Fig. 17. Dorsal fin single but almost separated into two fins (northern largemouth bass, *Micropterus s. salmoides*).

KEY TO THE FAMILIES OF GREAT LAKES FISHES

This family key will not apply consistently to regions distant from the Great Lakes. The characters used to separate families in this drainage basin do not necessarily characterize them throughout their entire range. Technicalities have not been entirely avoided. We are reminded that George Baker, to justify his use of technical terms, wrote, in the time of Shakespeare, "I would not have an ignorant asse be made a chirurgeon by my booke." This wise statement might be paraphrased to apply to a twentieth-century student of ichthyology.

1A Mouth a sucking disc (Figs. 18 and 19, p. 36; gill-openings seven (indicated by arrow on following figure); paired fins absent; single median nostril (Marsipobranchii; Hyperoartia)

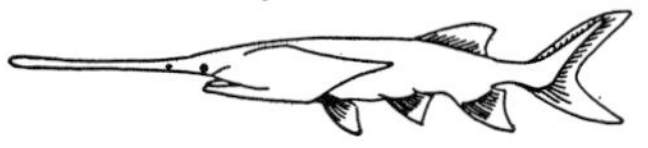

Lamprey Family
Petromyzontidae
p. 34

1B Mouth with true jaws; gill-slits four, covered by opercular bones; paired fins present; paired nostrils (Gnathostomi) 2

2A Caudal fin typically heterocercal (Fig. 8, p. 28); endoskelton largely cartilaginous (Chondrostei) 3

2B Caudal fin abbreviate-heterocercal (Fig. 9, p. 28) or nearly or quite homocercal (Fig. 10, p. 28) endoskeleton largely bony 4

3A Snout elongated and paddle-like (Fig. 20, p. 37); maxillary obsolete; skin of body naked

Paddlefish Family
Polyodontidae
p. 37

3B Snout relatively short and conical (Fig. 21, p. 38); maxillary developed; body partly plated

Sturgeon Family
Acipenseridae
p. 37

4A Caudal fin abbreviate-heterocercal (Fig. 9, p. 28) 5

4B Caudal fin nearly or quite homocercal (Fig. 10, p. 28) 6

5A Body covered with thick, rhombic ganoid scales; dorsal fin short; face produced into a beak (Fig. 22, p. 39)

Gar Family
Lepisosteidae
p. 39

5B Body covered with cycloid scales, (lacking ctenii, see Fig. 7, p. 23); dorsal fin elongate; no beak developed

Bowfin Family
Amiidae
p. 40

6A Dorsal fin with only one (Fig. 11, p. 28) or no principal spinous rays; pelvic fin without a spine (Malacopterygii) 7

6B Dorsal fin with more than one spine and pelvic fin with a spine (Acanthopterygii)* 21

* The spines are only weakly developed in some groups, such as the troutperch, sculpin and silverside families, and require magnification and dissection to observe. See Figs. 14 and 50, pp. 28 and 118.

7A Chin with one or more barbels (see next two figures in this key) 8

7B Chin without barbels 9

8A Chin with several barbels; pectoral and dorsal each with a hard spinous ray; scaleless

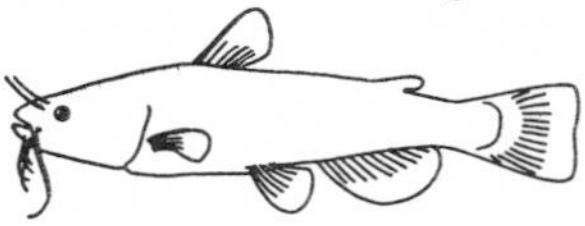

North American Catfish Family
Ictaluridae
p. 87

8B Chin with a single barbel (indicated by arrow on following figure); no spinous rays in pectoral and dorsal fins; small cycloid scales present (Anacanthini)

Cod Family
Gadidae
p.98

9A Body eel-shaped; dorsal and anal fins continuous with caudal fin (Apodes)

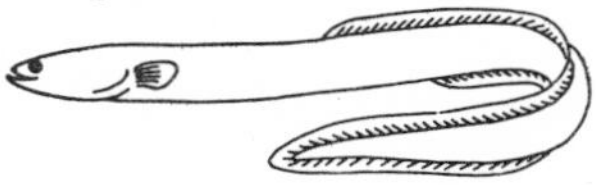

Freshwater Eel Family
Anguillidae
p. 95

9B Body not eel-shaped; dorsal and anal fins not continuous with caudal fin 10

10A Caudal fin rounded 11

10B Caudal fin not rounded 13

11A Most of the length of the pelvic fins behind origin of dorsal fin; premaxillaries not protractile (bound to snout by a fleshy frenum) (Haplomi, part)

Mudminnow Family
Umbridae
p. 92

11B Most (or all) of length of pelvic fins lying in advance of origin of dorsal fin; premaxillaries protractile (separated from snout by a groove) (Cyprinodontes) 12

12A Anal fin of male like that of female; third anal ray (counting rudiments) branched; egg-layers

Killifish Family
Cyprinodontidae
p. 95

12B Anal fin of male not like that of female, modified into an intromittent organ (indicated by arrow on following figure); third anal ray unbranched; live-bearers

Livebearer Family
Poeciliidae
p. 97

13A Gill-slits extended far forward below; gill-membranes free from isthmus (Fig. 12, p. 28) 14

13B Gill-slits not extended far forward below; gill-membranes united to isthmus (Fig. 13, p. 28) 20

14A No adipose fin developed 15

14B An adipose fin developed (indicated by arrow on figure given below, under item 17A) (Isospondyli; Salmonoidea) 17

15A Midline of belly with strong spiny scutes forming a saw-like keel (indicated by arrows on following two figures); no lateral line on body (Isospondyli; Clupeoidea)

Herring Family
Clupeidae
p. 42

15B Midline of body without strong spiny scutes; lateral line present on body, although not always continuous 16

16A Front of head shaped like a duck's bill; scales developed on head; jaws with strong teeth (Fig. 39, p. 93) (Haplomi, part)

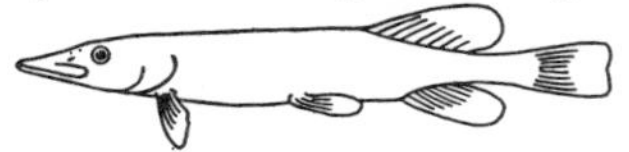

Pike Family
Esocidae
p. 92

16B Front of head not shaped like a duck's bill; no scales on head; jaws without strong teeth (Isospondyli; Notopteroidea)

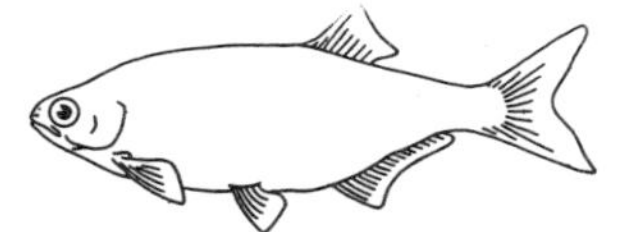

Mooneye Family
Hiodontidae
p. 41

17A Scales small, more than 100 in lateral line

Salmon Family
Salmonidae
p. 43

17B Scales larger, fewer than 100 in lateral line 18

18A Length of depressed dorsal fin longer than head; dorsal fin with more than 15 rays

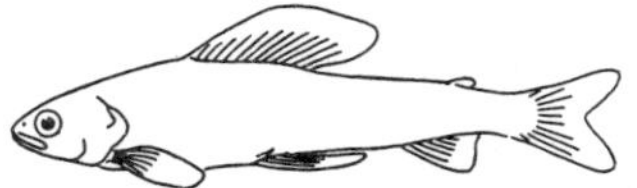

Grayling Family
Thymallidae
p. 56

18B Length of depressed dorsal fin shorter than head; dorsal fin with fewer than 15 rays 19

19A Teeth strong; maxillaries (upper jaw bones) extending behind middle of eye; no pelvic axillary process

Smelt Family
Osmeridae
p. 57

19B Teeth very weak or absent; maxillaries not extending behind middle of eye; a pelvic axillary process (indicated by arrows on the following two figures)

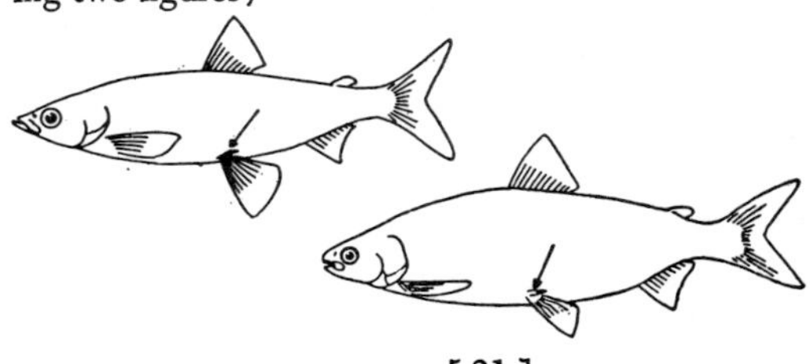

Whitefish Family
Coregonidae
p. 48

20A Mouth fitted for sucking (Fig. 33, p. 61); distance from front of anal fin to base of caudal fin contained more than 2.5 times in distance from front of anal fin to tip of snout (the carp and goldfish, members of the minnow family, will enter this category but may be told from the suckers by the strong hardened rays at the front of the dorsal and anal fins, Figs. 11, p. 28 and 110–112); pharyngeal teeth numerous and usually comb like, in a single row (Figs. 15, p. 28, and 32, p. 60)

Sucker Family
Catostomidae
p. 57

20B Mouth not fitted for sucking; distance from front of anal fin to base of caudal fin contained less than 2.5 times in distance from front of anal fin to tip of snout; pharyngeal teeth fewer than nine on each side and in one to three rows (Fig. 36B, p. 69)

Minnow Family
Cyprinidae
p. 66

21A An adipose fin developed; only two dorsal spines

Troutperch Family
Percopsidae
p. 98

21B No adipose fin; more than two dorsal spines 22

22A Anus far in advance of anal fin, located near throat (position indicated by arrow on following figure)

Pirateperch Family
Aphredoderidae
p. 99

22B Anus located posteriorly, immediately before anal fin 23

23A Dorsal spines isolated, not connected to one another by membranes

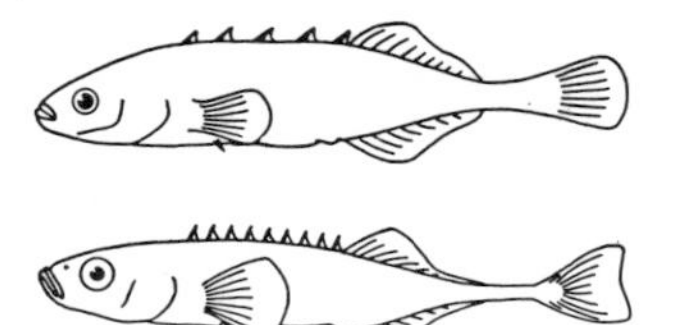

Stickleback Family
Gasterosteidae
p. 119

23B Dorsal spines not isolated, connected to one another by membranes (Figs. 16 and 17, p. 28) 24

24A Body scaleless, or with prickles only; pelvic fin (Fig. 50, p. 118) with only three or four soft-rays; no anal spines

Sculpin Family
Cottidae
p. 116

24B Body covered with ordinary scales; pelvic fin with five soft-rays; one or more anal spines (Percomorphi) 25

25A Lateral line extended across caudal fin (indicated by arrow on following figure)

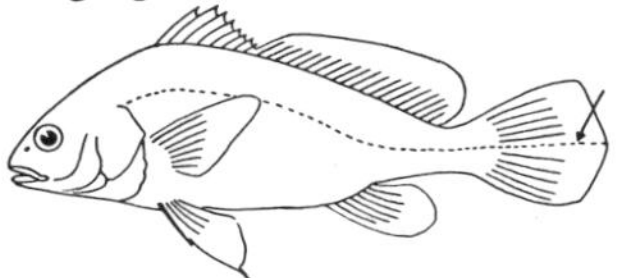

Drum Family
Sciaenidae
p. 116

25B Lateral line not extended across caudal fin 26

26A Pelvic fins abdominal, located far behind pectoral fins; origin of anal fin in front of origin of dorsal

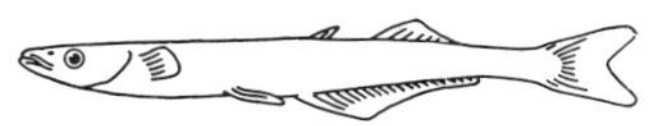

Silverside Family
Atherinidae
p. 115

26B Pelvic fins thoracic, located close to pectoral fins; origin of anal fin behind that of dorsal 27

27A Dorsal fin single (Fig. 16 p. 28) (almost separated in largemouth bass, Fig. 17, p. 28)

Sunfish Family
Centrarchidae
p. 110

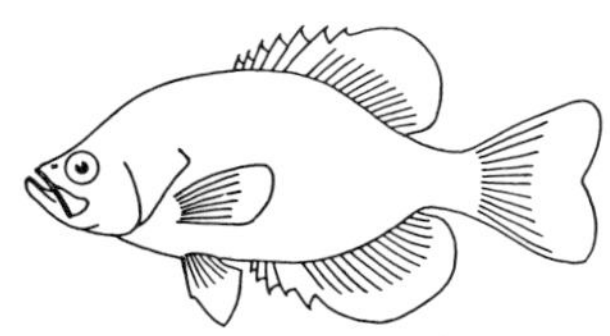

27B Dorsal fins two, entirely separated or but slightly joined together 28

28A Anal spines three

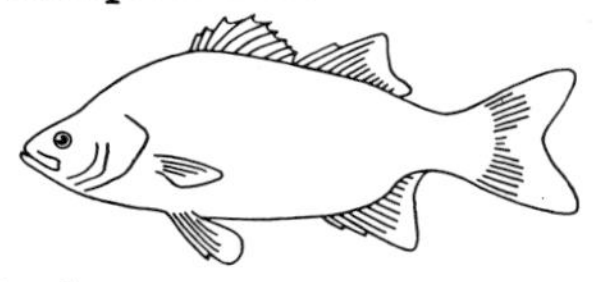

Bass Family
Serranidae
p. 99

28B Anal spines one or two

Perch Family
Percidae
p. 100

THE FAMILIES OF GREAT LAKES FISHES, WITH KEYS AND SPECIES ACCOUNTS

For each of the twenty-nine families of fishes represented in the Great Lakes basin, materials are presented as follows. First, there is an account of the family, indicating recognizable characteristics, a statement of geographical distribution, and, in condensed form, knowledge regarding habits and relations to man. Second, there is a key for the identification of the species and subspecies recognized for the region. Where required, particularly difficult key characters are illustrated. The keys are adequate for adult fish, only sometimes for juvenile ones, and, as indicated before, often only for the family within the Great Lakes region. A special key to larvae is provided to enable the identification of sea lamprey ammocetes, because of the extreme, current significance of the species in the Great Lakes. Following the keys there is a word-sketch of the range of each species and subspecies; known limits are given. When only one subspecies is recognized for the Great Lakes region, the total range of the species is indicated approximately by a statement in the account of that subspecies.

LAMPREY FAMILY—*Petromyzontidae*

(Figs. 51–55)

The lampreys are characterized by their eel-like body shape, the sucking-disc mouth, lacking the ordinary upper and lower jaws, the absence of paired (pectoral and pelvic) fins, and the seven external gill openings on each side.

About two dozen species occur in the range of the family, which includes North America from the Arctic to Mexico and also Eurasia and the Subantarctic regions. The group is particularly well represented in eastern North America.

Five kinds have been recorded in the Great Lakes and tributary waters. Of these the sea lamprey alone is not strictly a freshwater form. It was formerly native in the Lake Ontario basin but has gained entry to the upper lakes during recent times through a canal, as previously indicated. There is some confusion of lampreys with the eel, and lampreys are often misnamed "lamprey eels." The true eel, of the family Anguillidae, has true jaws which are lacking in the lampreys.

Lampreys inhabit creeks, rivers and lakes. These creatures all ascend streams to spawn in the spring. They ordinarily nest on gravelly riffles, where they dig the shallow pits within which they subsequently spawn. After several days the young appear and drift downstream until they become lodged in the mud of a bank in some quiet back-water. Here they burrow into the bottom material and spend several years as larvae, feeding on materials which they strain from the oozy layer on the bottom. In due time when a size of a few inches has been reached (varying with the species), the larvae, called ammocetes, transform during late summer and fall into an adult-like lamprey. They lose the fleshy hood which overhangs the mouth and gain the true sucking disc set with variable teeth. At transformation they assume habits that provide a basis for classifying the lampreys into two types. One type, which becomes parasitic on other fishes, is represented by the silver, chestnut and sea lampreys. The other type is represented by the two brook lampreys—the northern and the American. In the latter, nonparasitic ones, the digestive tract becomes degenerate and the metamorphosed individual does not feed but merely lives until the next spawning season, when it reproduces and dies. In the parasitic forms, the digestive tract remains functional. These lampreys attach themselves to fishes and, after rasping a hole through the body covering, suck nourishment from the host. Following a period of parasitization (usually a single growing season) they attain sexual maturity, spawn and die.

The nonparasitic lampreys stop growing in length before the time of metamorphosis, at lengths less than ten inches. Most native parasitic species reach the length of about one foot. The sea lamprey may become nearly three feet long but in the Great Lakes ordinarily does not exceed two feet, and averages between seventeen and eighteen inches.

Lampreys have some value as study specimens in schools, for they represent a very primitive group of fishes. Among living animals, the larval lamprey is regarded as most like the prototype of the vertebrates. Adult lampreys are sometimes eaten and are reported to make very excellent food and the larvae are used as bait. Great damage is done by parasitic lampreys to Great Lakes fishes of both sport and commercial significance. The problem of controlling the sea lamprey in the upper Great Lakes remains unsolved. Indications are that control of spawners by weirs across streams and destruction of larvae by selective poisons may work.

KEY TO ADULTS

1. Dorsal fin continuous, never divided into two distinct fins (Figs. 51-53); trunk myomeres (muscle segments between last gill opening and anterior edge of vent slit) 47 to 58 (*Ichthyomyzon*) 2
   Two distinct dorsal fins either close together or well separated (Figs. 54, 55); trunk myomeres 64 to 74 4

2. Nonparasitic species; adults with nonfunctional intestinal tract reduced to a mere strand of tissue; teeth and buccal funnel degenerate.
   NORTHERN BROOK LAMPREY—*Ichthyomyzon fossor* Reighard and Cummins. (Fig. 52)
   Parasitic species; adults retaining a functional digestive tract with a wide lumen; teeth and buccal funnel well developed 3

3. Circumoral teeth (with very rare exceptions) all unicuspid; trunk myomeres usually 49 to 52.
   SILVERY LAMPREY—*Ichthyomyzon unicuspis* Hubbs and Trautman. (Fig. 51)
   Circumoral teeth at least in part bicuspid (with occasional exceptions); trunk myomeres usually 51 to 54.
   CHESTNUT LAMPREY—*Ichthyomyzon castaneus* Girard. (Fig. 53)

4. Buccal funnel with a series of teeth radiating in all directions from the mouth (Fig. 18).
   SEA LAMPREY—*Petromyzon marinus* Linnaeus. (Fig. 54)
   Buccal funnel with teeth not in radiating series, but in several groups (Fig. 19).
   AMERICAN BROOK LAMPREY—*Entosphenus lamottenii lamottenii* (LeSueur). (Fig. 55)

KEY TO LARVAE WITH TWO DORSAL FINS
(Characters from Vladykov, 1950)

Trunk myomeres averaging 70 (range, 67-74); outline of tail fin rounded (like a beaver's tail); pigmentation of caudal fin rays extending to the margins of the rays; lower limits of caudal myomeres with grayish chromatophores; pigmentation in branchial region extending downward from back nearly to branchial groove; suborbital area well pigmented and lower half of lip at least partly so.
SEA LAMPREY—*Petromyzon marinus* Linnaeus.

Trunk myomeres averaging 67 (range 63-70); tail fin bluntly pointed; pigmentation of caudal fin rays confined to their bases; lower limits of caudal myomeres unpigmented; pigmentation in branchial region extending downward from back only half the distance to branchial groove; suborbital area and lower half of upper lip unpigmented.
AMERICAN BROOK LAMPREY—*Entosphenus lamottenii lamottenii* (LeSueur).

Silvery lamprey–*Ichthyomyzon unicuspis* Hubbs and Trautman. Fig. 51.—Interior regions from the Ohio River basin including Kentucky through the eastern part of the upper Mississippi Valley (including Missouri, eastern Iowa, and southeastern Wisconsin) to the western tributaries of Hudson Bay; basins of lakes Superior, Michigan (chiefly in the main lake and bays), Huron, Erie and Ontario (rare and local in Lake Ontario); the St. Lawrence River drainage downstream to Montmagny, Quebec, and Lake Champlain. Generally in larger lakes and rivers. Parasitic as adult.

Northern brook lamprey–*Ichthyomyzon fossor* Reighard and Cummins. Fig. 52.—Mississippi River drainage in Wisconsin, northeastern Illinois, and northern Indiana, from all of the Great Lakes basins in Michigan and southern Ontario to a Lake Erie tributary in New York and to southern, upper St. Lawrence River tributaries in Quebec. In creeks and small rivers, living for several years as ammocetes in sand mixed with organic matter, transforming in late summer or fall, passing the winter without food and spawning the next spring on stony or gravelly riffles.

Fig. 18. Fig. 19.

Fig. 18. Buccal funnel of the sea lamprey, *Petromyzon marinus.* After Gage, 1893.)

Fig. 19. Buccal funnel of the American brook lamprey, *Entosphenus lamottenii lamottenii.* (After Gage, 1893.)

Chestnut lamprey–*Ichthyomyzon castaneus* Girard. Fig. 53.—Interior lowlands from western Manitoba, Iowa, and northern Wisconsin to the Alabama River system in Georgia and to Louisiana, northeastern Texas, and eastern Oklahoma; generally replaced by *I. bdellium* in the upper Ohio Valley but ascending the Tennessee River to Alabama. In the Great Lakes confined almost wholly to Lake Michigan and all its main tributaries (penetrating the Lake Huron drainage area only to the Cheboygan River). Ammocetes living several years in peaty sand; adults parasitic on fishes for one year.

Sea lamprey–*Petromyzon marinus* Linnaeus. Figs. 18 and 54.–North Atlantic Ocean from Iceland and northern Europe to northwestern Africa, and from the Grand Banks and the Gulf of St. Lawrence (and perhaps from Labrador) to northern Florida. More or less landlocked in Lake Champlain and Lake Ontario (and tributary lakes), from which it has recently spread through canals to become well established in all the other Great Lakes. Larval life of three or more years spent in silty bottoms of streams. Metamorphosing in streams to adult stage. Then parasitic on larger fishes for about 14 to 18 months in lakes until spawning run to streams begins in spring.

American brook lamprey–*Entosphenus lamottenii lamottenii* (LeSueur). Figs. 19 and 55.—In the Mississippi River drainage from southern and eastern Minnesota to western Pennsylvania, south to Tennessee and Missouri; throughout all the Great Lakes basins and down the St. Lawrence River and its tributaries, including Lake Champlain, to Montmagny, Quebec; on the Atlantic slope from the Connecticut and Hudson river systems to Maryland. The same or a virtually indistinguishable form is in the Yukon River system in Alaska and in northeastern Asia. In Alaska it appears to intergrade with the typical, often anadromous parasitic form *japonicus* that ranges from north-central Canada to the White Sea and south to Japan and Korea. Mostly in creeks, living several years in sand mixed with some organic matter and then metamorphosing, in the fall, to spawn the following spring on the fine gravel of shallow riffles.

Paddlefish Family—*Polyodontidae*

(Figs. 20 and 56)

The paddlefish is distinguished by its elongated, paddle-shaped snout (Fig. 20). In common with the sturgeon, it has a strongly upturned tail. Furthermore, scales are lacking except on part of the tail.

The only known living near-relative of the paddlefish is a gaint fish, *Psephurus,* of the great rivers of China, and only two or three fossil paddlefishes are known. They are very primitive fishes, remarkably shark-like in some respects.

This fish has been very rare and is perhaps now extinct in the Great Lakes. It is also much depleted in the northern part of its Mississippi Valley range. The paddlefish is essentially an inhabitant of the open waters of large, silty rivers and of oxbow and flood-plain lakes. Little is known about the spawning habits of this fish or about its early life history. It obtains food by straining microscopic organisms from the water through the efficient sieve-like gill-rakers. It grows to large sizes, exceeding six feet in length and 150 pounds in weight.

In its former period of abundance in the Mississippi Valley, it had considerable market value. Its flesh was used for food and its roe was made into caviar.

This fish is of particular biological interest in that it represents one of the more primitive groups of fishes and has a chiefly cartilaginous skeleton.

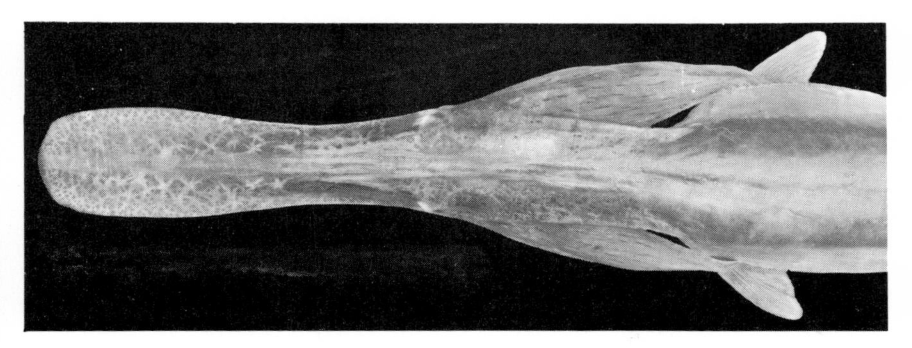

Fig. 20. Head of paddlefish (*Polyodon spathula*) viewed from above.

Paddlefish—*Polyodon spathula* (Walbaum). Figs. 20 and 56.—Mississippi River system from the Missouri River, in eastern Montana, to Pennsylvania and New York (Lake Chautauqua); southward to western North Carolina, Mississippi, Louisiana and Texas. Recorded a very few times from the Great Lakes basin, and thought by some to have reached these waters *via* canals, but more likely the species was encountered on the way to natural extirpation in the north; the Great Lakes reports (all dating from more than 50 years ago) are for Lake Erie, Spanish River on Georgian Bay, Lake Huron near Sarnia, Lake Winnebago or Green Bay (prior to 1700), Lake Helen on the Nipigon River, and, doubtfully, Lake Ontario. Most often in the open waters of large, silty rivers.

Sturgeon Family—*Acipenseridae*

(Figs. 21 and 57 and col. pl. no. 1)

The sturgeons, of which only the lake sturgeon is present in the Great Lakes, are marked by the rows of bony scales which partially armor the body. The tail fin is heterocercal. The protrusible mouth is situated ventrally well behind the tip of the conical snout (Fig. 21), which bears two pairs of barbels that are usually in contact with the bottom.

Only a few kinds of sturgeon remain, to represent in degenerate form a group that was dominant in early geological history. The living species are confined to the northern parts of North America and Eurasia.

The lake sturgeon was formerly rather abundant throughout most of the Great Lakes region, where it frequented the shoal waters of the larger lakes and also at times some of the larger rivers. It attained its greatest abundance in Lake Erie and its least abundance in Lake Superior. It became greatly reduced in its numbers in earlier days. Protective legislation is apparently bringing about some come-back of this species at the present time.

Spawning takes place in the late spring and in the early summer in tributary streams and perhaps also in the shallow waters of the Great Lakes. The rate of growth is slow but large sizes are attained. Sexual maturity supposedly does not come until after twenty years of life. Weights attained exceed 300 pounds and lengths, seven feet. As one might expect from the position of the toothless mouth, the sturgeon is a bottom feeder. It sucks up bottom material and strains out the food organisms.

The sturgeon is a very desirable food fish, for its flesh is tasty, white and flaky and its roe may be made into delicious caviar. It is of biological interest because it represents a very primitive type of fish.

Fig. 21. Lower surface of head of lake sturgeon (*Acipenser fulvescens*), showing ventral position of mouth and tactile barbels under snout. (Courtesy Institute for Fisheries Research, Michigan Department of Conservation.)

LAKE STURGEON—*Acipenser fulvescens* Rafinesque. Figs. 21 and 57 and color plate no. 1.—From the Red River of the North, the Saskatchewan River in Alberta, Hudson Bay tributaries from the Churchill River southward, and the St. Lawrence and Lake Champlain drainages of Canada (and possibly from Labrador), southward, west of the Appalachian Mountains, to the Tennessee River of Alabama, to Missouri, to eastern Nebraska, and to northeastern Kansas. Typically in larger rivers and lakes, ascending medium-sized streams to spawn on riffles. Now greatly reduced in numbers in the Great Lakes and throughout almost its entire range.

Gar Family—*Lepisosteidae*

Gars are distinguished by the more or less diamond-shaped thick ganoid scales that cover the body. Furthermore, the jaws and face are extended forward into a beak. The teeth are exceptionally strong, sharp and conical (Fig. 22.).

Fig. 22. Head of adult northern longnose gar (*Lepisosteus osseus oxyurus*), showing beak-like form and the numerous sharp teeth.

(Courtesy Institute for Fisheries Research, Michigan Department of Conservation.)

This family is exclusively North American, ranging southward to Cuba and Central America. There are fewer than ten species, all freshwater. Gars are spoken of as "living fossils" because nearly all their relatives are extinct.

The gars are represented in Great Lakes waters by two widely ranging species. Of these, the longnose gar is the more abundant in this fauna. These fishes frequent quiet, warm waters in lakes and larger streams. They are sluggish in their habits except when feeding, at which time they move swiftly to capture their prey, mostly composed of other fish. It is a common habit of the gars to bask near the surface on warm days or nights where they may be seen floating like logs of varying sizes.

Spawning takes place in the shallows in the spring. Growth is quite rapid, at least in early lfe. The longnose gar commonly attains a length of five feet; the spotted gar, a length of three feet.

Under certain circumstances, these predacious fishes become so abundant that the problem of their control arises. It is common practice in several states to remove gars from fishing waters by gill netting and also by regulated spearing on quiet nights, with the aid of a jacklight from slowly moving boats. Gars are sometimes nuisance fish to anglers fishing with live bait since they steal the bait from hooks but are not themselves easily caught in this way. They may, however, be taken by using a piano-wire snare onto which a minnow is threaded. The loop of this snare may be drawn tight about the snout of a gar when it strikes the bait.

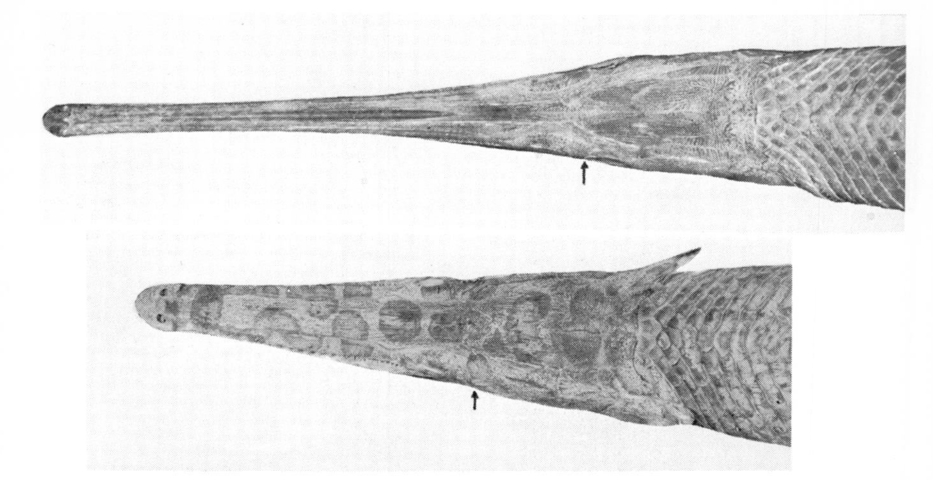

Fig. 23. Dorsal views of heads of two adult gars to show differences in proportions and coloration. Upper figure: northern longnose gar (*Lepisosteus osseus oxyurus*). Lower figure: spotted gar (*L. productus*); note dark spots on top of head and beak. Arrows indicate end of beak (at anterior rim of orbit). Note the difference in relative length and width of beaks, as stated in the key.

Beak shorter and broader, its length less than 10× its least width (this proportion varies with size of fish).
SPOTTED GAR—*Lepisosteus productus* (Cope). (Figs. 23, 58)
Beak long and slender, its length more than 12× its least width (except in young; varying with size of fish).
NORTHERN LONGNOSE GAR—*Lepisosteus osseus oxyurus* Rafinesque. (Figs. 22, 23, 59)

SPOTTED GAR—*Lepisosteus productus* (Cope). Figs. 23 and 58.—From the Lake Michigan drainage of Michigan, from Saginaw Bay of Lake Huron, and from various parts of the Lake Erie basin, including that in southern Ontario, to the upper Mississippi River system in eastern Missouri and Illinois, and possibly to Chautauqua Lake, New York; southward (occasionally even in brackish water) to western Florida and the Rio Grande. Northward almost wholly confined to clear, weedy, glacial lakes; southward most frequently in weedy bayous, as contrasted to the open, silty rivers in which the shortnose gar, *L. platostomus,* abounds.

NORTHERN LONGNOSE GAR—*Lepisosteus osseus oxyurus* Rafinesque. Figs. 22, 23 and 59.—From Mississippi River affluents in Montana through the Great Lakes basins (apparently excepting that of Lake Superior) to the St. Lawrence-Champlain watershed of Quebec and Vermont, descending the St. Lawrence River, as elsewhere in its range, to salt water; south to northern Alabama and the highlands of northern Mexico. *L. o. osseus* (Linnaeus) lives in waters of the extreme southeastern United States, including peninsular Florida, to complete the range of the species. Usually near the surface in open rivers and lakes; northward most commonly in weedy glacial lakes.

BOWFIN FAMILY—*Amiidae*

(Fig. 60 and col. pl. no. 2)

The bowfin may be told from all other fishes of the region by the long fin that arches in a bow over most of the length of the back of the fish. The body is covered with cycloid scales and there is a unique bony plate on the undersurface of the lower jaw at the front, just where the two sides meet.

Only one species of this family is extant. Like the gars, it is termed a "living fossil" because all its numerous relatives have long since become extinct.

The bowfin, like the gars, is a quiet-water inhabitant and reaches the northward limits of its distribution in the Great Lakes region. Spawning takes place in the spring in the shallows and usually in vegetation, where the male clears a circular area for a nest. The eggs are subsequently guarded by the male who also herds the young about for a time after hatching.

Little is known about the age and growth of the bowfin since the scales apparently do not lend themselves to age determination by the reading of year marks, as do the scales of many other fishes. Ordinary size attained by the females may be as much as three feet. The males are smaller. In the adult, sex may be distinguished by the fact that the spot (ocellus) at the upper base of the tail fin in the male is rimmed with orange-yellow, whereas in the females this rim is lacking or the spot is obsolete. The food of the bowfin is mostly composed of crayfish and fish. Some game and pan fishes are of course eaten. This fact, coupled with the abundance that the species attains in certain waters, since it is not sought after by the angler, has caused it to be regarded as an obnoxious, predatory fish. In this respect, some control has been practiced; the means used have been mostly netting and spearing. Although the fish is eaten in the South, it has only a small food or market value in the Great Lakes region.

The young are often confused with mudminnows by bait dealers. Fishermen should therefore use caution in the release of any fish of uncertain identity or of questionable relationships.

BOWFIN—*Amia calva* Linnaeus. Fig. 60 and color plate no. 2.—From the Mississippi River system in Minnesota through Lake Nipissing and the Ottawa River to the St. Lawrence-Champlain basin in Quebec and Vermont; southward, west of the Appalachians, to Florida, southeastern Oklahoma, and Texas; northward on the Atlantic slope to the Carolinas and to the Susquehanna River (recorded from Connecticut, probably the result of introduction). Throughout the Great Lakes region, though not in the drainage basin of Lake Superior (except its outlet, St. Marys River).

MOONEYE FAMILY—*Hiodontidae*

(Fig. 61 and col. pl. no. 3)

The mooneye superficially resembles a true herring (family Clupeidae) but it is distinguished by the lack of a row of sharp, spiny scutes down the midline of the belly and by the large eye. Furthermore, the scales are not like those of the herring.

There are only two or three species in this family. They occur only in the fresh waters of eastern North America and constitute another primitive element in that fauna.

The mooneye is the sole representative of its family in the Great Lakes basin. It is found in the open waters of larger lakes and streams. Very little is known about its life history. Its food is reported to be composed principally of insects and their larvae and small minnows. Occasionally mooneyes are taken by anglers and the species enters the commercial catch in a very small way. It is not highly valued as food.

In the northern and southwestern parts of its range, but apparently not in the Great Lakes, the mooneye is accompanied by the goldeye (*Amphiodon alosoides*), which differs in having the dorsal fin beginning farther back than the anal and the fleshy keel on the belly extending forward beyond the pelvic fins.

MOONEYE—*Hiodon tergisus* LeSueur. Fig. 61 and color plate no. 3.—From the western and southern tributaries of Hudson Bay, including Lake Winnipeg, to the St. Lawrence-Champlain drainage; south through western Pennsylvania and western Maryland to northern Alabama, southern Arkansas, and southeastern Oklahoma; west to the eastern parts of Kansas and to Manitoba and Saskatchewan. In the Great Lakes confined to the southern part of the basin, including lakes St. Clair, Erie and Ontario.

Herring Family—*Clupeidae*

(Figs. 62–64 and col. pl. no. 4)

Members of the herring family are characterized by having a strong, sharp-edged row of spiny scutes along the midline of the belly which give it a saw-tooth appearance. The kind of cycloid scales that they possess is further characteristic. The ridges (circuli) on these scales run more or less across the scale from upper edge to lower edge rather than being more or less concentrically arranged in rings or running longitudinally. In general, these are silvery fishes with very slab-sided bodies.

Fishes of this family abound in the oceans. Some, like the Atlantic herring and the Pacific sardine, are of enormous economic importance. Most of the numerous species are marine, but some are anadromous and some are freshwater. The group is best developed in Tropical waters, but some approach the Polar seas.

The herrings of the Great Lakes region are essentially dwellers of the large lakes. The gizzard shad is common in the southern portion of the drainage, whereas the other two forms, alewife and American shad, are confined principally to Lake Ontario. The alewife has now penetrated all of the upper Great Lakes, presumably having gained access to them through the Welland Canal.

These fishes are of some significance in our fauna as forage organisms for other fish. They afford here no commercial fishery. In Lake Ontario, particularly along the south side, the alewife is a nuisance in that large numbers die each spring and wash ashore to make a noisome litter for cottagers and resort owners. The reasons for this annual mortality are not fully known.

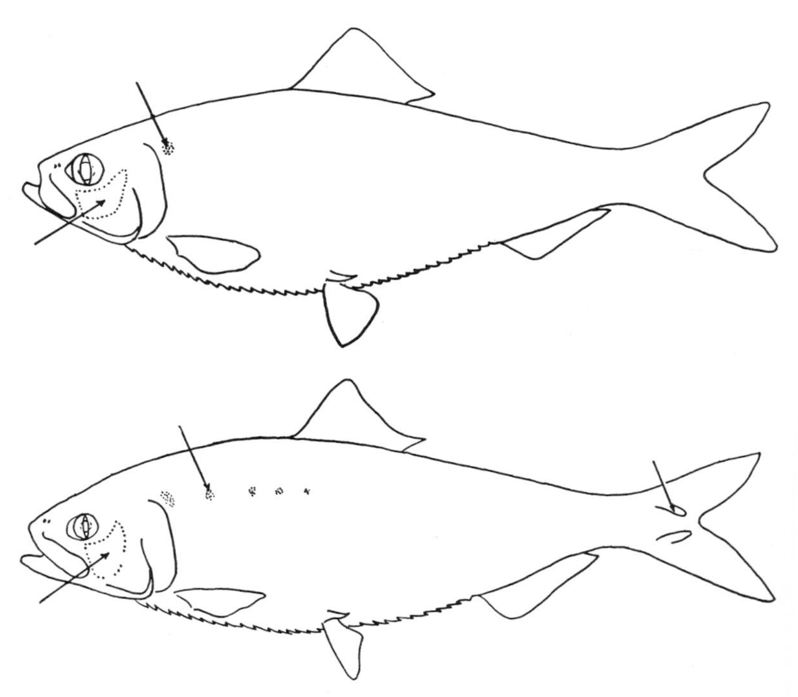

Fig. 24. Features distinguishing the alewife (*Pomolobus pseudoharengus*), above, from the American shad (*Alosa sapidissima*), below. Note differences in dimensions of silvery area on cheek (dotted outline), number of spots behind upper corner of gill cover, and the presence of the alar scales near the caudal-fin base in the shad. (Modified from Legendre, 1954).

1
- Last dorsal ray greatly elongated; stomach gizzard-like; snout thick, lower jaw included.
  GIZZARD SHAD—*Dorosoma cepedianum* (LeSueur). (Fig. 64, col. pl. no. 4)
- Last dorsal ray not prolonged; stomach not gizzard-like; snout sharp, lower jaw more or less projecting 2

2
- Silvery area of cheeks deeper than long; more than 55 gill-rakers on lower part of anterior arch; upper jaw terminal; two flaps of skin, each covered with an elongated alar scale, at base of caudal fin (Fig. 24); three or more dark spots on side of body behind upper end of gill opening.
  AMERICAN SHAD—*Alosa sapidissima* (Wilson). (Fig. 63)
- Silvery area of cheeks longer than deep (Fig. 24); fewer than 55 gill-rakers on lower part of anterior arch; lower jaw terminal; no flaps of skin at base of caudal fin; one dark shoulder-spot on side of body behind upper end of gill opening.
  ALEWIFE—*Pomolobus pseudoharengus* (Wilson). (Fig. 62)

ALEWIFE*—*Pomolobus pseudoharengus* (Wilson). Fig. 62.—Atlantic Coast from Labrador and the St. Lawrence River drainage in Ontario and Quebec to Florida. Anadromous. Originally landlocked in Lake Ontario and some tributary lakes (where possibly introduced); recently (1954) having completed a spread through all of the upper Great Lakes (including depths to 30 fathoms).

AMERICAN SHAD—*Alosa sapidissima* (Wilson). Fig. 63.—Atlantic Coast from Labrador to Florida, developing in the sea but ascending streams to spawn. In the Great Lakes rare in the Lake Ontario basin (no recent records). In Ontario now occurring only in the Ottawa River. Once taken in Lake Erie as a result of stocking. Introduced into the Pacific where it now ranges from southern California to southern Alaska.

GIZZARD SHAD—*Dorosoma cepedianum* (LeSueur). Fig. 64 and color plate no. 4.—From Nebraska and Minnesota to the St. Lawrence River (down to Quebec City) and to the Ohio Valley in western Pennsylvania; south to the Gulf of Mexico and to eastern Mexico; coastwise north to the middle Atlantic coast in New Jersey and Pennsylvania. In the Great Lakes common in the southern portion of the drainage and occasionally northward to Georgian and Saginaw bays. Ordinarily in bays, estuaries, lakes, bayous and large rivers, in clear to very silty water.

SALMON FAMILY—*Salmonidae*

(Figs. 65–72, col. pl. p. xiv and col. pl. no. 5)

Salmon and trout are fine-scaled fishes, having more than a hundred scales in the lateral line. They have pelvic axillary processes and an adipose dorsal fin. The jaws and other bones carry moderately well developed teeth. Comparisons with other families are made on pp. 48 and 56. Some workers include the fishes of the family Coregonidae in the Salmonidae.

As here restricted the family Salmonidae includes only Arctic and North Temperate fishes—salmons, trouts and chars—that either live throughout their lives in fresh water or grow in the sea but run into streams to spawn. The species are relatively few, but there is no agreement among authorities on the total number.

The principal members of the salmon family of the Great Lakes region are the brook, brown, rainbow and lake trouts. The landlocked salmon, Yellowstone cutthroat, chinook salmon and Cascapedia trout have all been introduced

* The inclusion of the Great Lakes in the range of the skipjack herring, *Pomolobus chrysochloris* Rafinesque, rests so far as we can ascertain on the report of the "saw-belly" by Milner (1872). But in his list of Great Lakes fishes Milner did not include *Dorosoma cepedianum*, now common in Lake Erie. In the absence of contrary evidence we may assume that his "saw-belly" refers to the gizzard shad, which is still generally known by that name in Lake Erie. Probably Milner's record was based on reports by fishermen.

or found in the basin at one time or another but are now either very rare or extinct. The brook trout and three forms of the lake trout are native to the area, whereas successful exotics are the brown trout, introduced from Europe, and the rainbow trout, brought in from the Pacific Coast. Local names applied to some of these trout may be confusing. For example, a rainbow trout that has run into the Great Lakes and developed into a silvery fish with a steely blue colored head and back is called a "steelhead," and sometimes even a "salmon" by an over-enthusiastic angler. Lakerun brown trout in the Great Lakes respond similarly colorwise and are easily confused with the landlocked Atlantic salmon. Brook trout that work along the shores of some of the lakes are called "coasters." "Mackinaw trout" are lake trout, *Salvelinus namaycush.*

Brooks, browns and rainbows are principally cool stream dwellers, whereas the common lake trout and the related siscowet and Rush Lake trout are lake inhabitants. Sometimes, however, the stream trouts find suitable abode in lakes with tributary trout streams or in landlocked lakes into which they have been introduced

Apparently all of the trouts do considerable migrating, particularly during the breeding season. The stream trouts characteristically spawn in clear, moving water of shallow depth on cleaned gravel. The eggs hatch under several inches of this bottom material and the fry make their way up to the open water. Lake trout spawn on rocky reefs. The food of smaller trout is almost entirely insects, whereas larger ones often add minnows and other items to their diet. Larger lake trout are almost entirely piscivorous. In general, trout spawning seasons are: lake, fall; brook, from very late summer to midwinter; brown, late fall to midwinter; rainbow, late winter to early spring depending on the latitude, on the stream and on the race of trout.

Many trout are raised each year at fish hatcheries and rearing stations for stocking to maintain the supply in fishing waters or for introduction into waters which they formerly did not inhabit. Increased research has brought about more judicious, less wasteful stocking, and has challenged the value of stocking the common lake trout in the Great Lakes proper.

The economic importance of trout in the Great Lakes region can hardly be overestimated. Lake trout have supported a large commercial fishery in the Lakes, other than Erie, and, in the Upper Lakes, also provide a sport fishery, "deep-sea trolling." Recently, however, the species has been greatly depleted. It appears that this is largely due to predation by the sea lamprey. Angling for trout, with its related recreational outlets of camping and boating, and with its needs for guides, gear and transporation, is one of the greatest assets of many northern communities of the area.

1. Anal fin longer than high, with 13 or more developed rays.
   CHINOOK SALMON—*Oncorhynchus tshawytscha* (Walbaum).

   Anal fin higher than long, with 12 or fewer developed rays 2

2. Vomer with a plane shaft bearing teeth in alternating rows or in one zigzag row (Figs. 25, 27) (posterior vomerine teeth sometimes few and deciduous); scales in fewer than 190 rows; species black- or brown-spotted, with reddish spots in the brown trout (*Salmo*) 3

   Vomer with trough-like, toothless shaft, and with teeth confined to head of bone (Fig. 26); scales in about 200 lateral rows; species spotted with gray or red (*Salvelinus*) 7

3. Vomerine teeth little developed, those on the shaft of the bone few and deciduous (*Salmo salar*) 4

Vomerine teeth well developed, those on the shaft of the bone numerous and persistent, arranged in one zigzag or two alternating rows (Figs. 25, 27) 5

4. Average for the least depth of the caudal peduncle not more than three-fourths distance from adipose to base of procurrent, rudimentary caudal rays and not more than two-thirds of the distance from the anal to caudal (except for parr).

SEARUN ATLANTIC SALMON—*Salmo salar salar* Linnaeus.

Average of least depth of caudal peduncle usually greater than three-fourths of the distance from the adipose to base of procurrent, rudimentary caudal rays and more than two-thirds of the distance from anal to caudal (except for parr).

LANDLOCKED ATLANTIC SALMON—*Salmo salar sebago* Girard. (Figs. 65, 66)

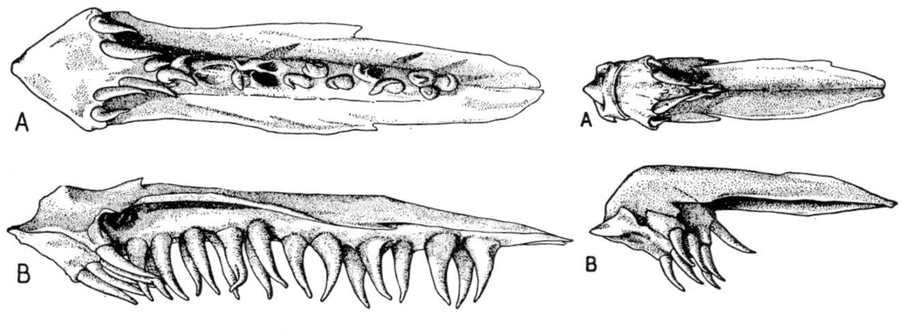

Fig. 25. Fig. 26.

Fig. 25 Vomer plane, with teeth on shaft (brown trout, *Salmo trutta fario*) A. Ventral view. B. Side view.

Fig. 26. Vomer without teeth on shaft, the teeth confined to the head of the bone (brook trout, *Salvelinus fontinalis*). A. Ventral view. B. Side view.

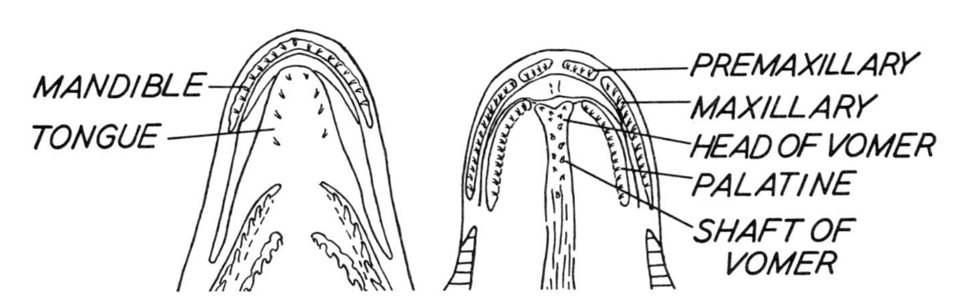

Fig. 27. Diagram of dentition in floor and roof of mouth of brown trout (*Salmo trutta fario*). The tooth-bearing structures or locations are labelled. Note teeth on both head and shaft of vomer; also shown on drawings of the isolated bone in Fig. 25.

5. A large, deep red or scarlet dash on each side concealed below the inner edge of each dentary bone ("cutthroat" mark), this rarely obsolete; small teeth usually present on hypobranchials of second gill-arch.
YELLOWSTONE CUTTHROAT—*Salmo clarkii lewisi* (Girard).

No red dash concealed below the inner edge of each dentary; no teeth on hpyobranchials of second gill-arch 6

6. Black (or brown) spots larger and more diffuse, scarcely developed on caudal fin; reddish spots more or less strongly developed, often ocellated with bluish; adipose fin of young orange, without dark margin or spots; pectoral fin larger and heavier.
BROWN TROUT—*Salmo trutta fario* Linnaeus. (Fig. 67 and col. frontispiece)

Black (or brown) spots smaller and sharper, well developed on caudal fin; red spots totally absent; adipose fin of young olive, with black margin or spots; pectorals shorter.
COAST RAINBOW TROUT—*Salmo gairdnerii irideus* Gibbons. (Figs. 68, 69)

7. Caudal fin little forked; body red-spotted; lower fins with a black stripe near leading edge; gill-rakers 9 to 12; precaudal vertebrae fewer than 37; mandibular pores (see Fig. 40, p. 94) usually 7 or 8; pyloric caeca fewer than 65; vomer without a long, raised crest, the teeth being confined to the head of the bone (Fig. 26).
BROOK TROUT—*Salvelinus fontinalis* (Mitchill). Figs. 70 and 71 and col. pl. no. 5)

Caudal fin rather strongly forked; body spotted with gray, without bright colors; lower fins without a black stripe; gill-rakers 12 to 14; precaudal vertebrae 38 to 41; mandibular pores (see Fig. 40, p. 94) usually 9 or 10; pyloric caeca more than 95; vomer with a raised crest extending backward from the head of the bone and free from the shaft, this crest armed with strong teeth (subgenus *Cristivomer; Salvelinus namaycush*) 8

8. Head nearly V-shaped as seen from above; fins rather short.
COMMON LAKE TROUT—*Salvelinus namaycush namaycush* (Walbaum). (Fig. 72)

Head rather broadly U-shaped as seen from above; fins larger 9

9. Size small; thin; confined to Rush Lake, northern Michigan.
RUSH LAKE TROUT—*Salvelinus namaycush huronicus* (Hubbs).

Size large; excessively fat; deep waters of Great Lakes.
SISCOWET—*Salvelinus namaycush siscowet* (Agassiz).

ATLANTIC SALMON—*Salmo salar* Linnaeus.—North Atlantic Ocean; in Europe south to Portugal, and in the western North Atlantic from southern Greenland, eastern Ungrava Bay, and Labrador south, originally to the Hudson River and occasionally to the Delaware River, but at present only to Maine. Landlocked in lakes of Quebec, the Maritime Provinces, and Maine. Native in the Lake Ontario basin but not found there since 1890. Introduced elsewhere.

The separate listing of the searun and landlocked forms is retained provisionally, though there is no assurance that they are genetically different, or, if so, that their characters justify such separation.

SEARUN ATLANTIC SALMON—*Salmo salar salar* Linnaeus.—Range of the species, inland only in waters accessible from the ocean. Native in the Lake Ontario basin but not found there since just before the turn of the century; still ascending nearby tributaries of the upper St. Lawrence River in Quebec to spawn in the fall.

LANDLOCKED ATLANTIC SALMON—*Salmo salar sebago* Girard. Figs. 65 and 66.—Populations in certain cold lakes of New England, Quebec, and the Maritimes have been classified as a separate subspecies. The separation is neither clear-cut nor of assured genetic basis. Similar landlocked forms occur in Europe. Landlocked Atlantic salmon from eastern waters have been stocked in the upper Great Lakes but are not known to have survived. Fishes resembling them in these waters (especially in western Lake Superior) are currently thought to be lakerun brown trout, *Salmo trutta fario*. In fact, it is difficult to separate adults of *S. salar* and *S. trutta*, both of which obviously belong in the same species group. Plantings in various inland lakes of Ontario, New York, and New England have been successful.

BROWN TROUT—*Salmo trutta fario* Linnaeus. Fig 67 and col. frontispiece. —Native in Europe and widely introduced in other regions, including much of the Great Lakes drainage area, where occasional in the Great Lakes proper and abundant in a few cool inland lakes and in many trout streams.

YELLOWSTONE CUTTHROAT—*Salmo clarkii lewisi* (Girard).—Headwaters of the Missouri River and middle and upper parts of the Columbia River system; represented coastwise from southern Alaska to northern California and in the Intermountain and Rocky Mountain regions by other subspecies of *S. clarkii;* introduced sporadically in the East (one recent record of at least temporary establishment in Michigan).

COAST RAINBOW TROUT—*Salmo gairdnerii irideus* Gibbons. Figs. 68 and 69. —Native in the Pacific Ocean from southern Alaska to southern California and perhaps northern Lower California; ascending coastwise streams to spawn. Now widely established elsewhere, including some cool lakes and many streams in nearly all parts of the Great Lakes watershed. Developing into the "steelhead" in the Great Lakes. Recent study of the type specimens fails to verify the subspecific distinctness of the "Shasta trout," of early fish culture, which was listed as *S. g. stonei* by Hubbs and Lagler (1941).

CHINOOK SALMON—*Oncorhynchus tshawytscha* (Walbaum).—North Pacific Ocean from Bering Sea to northern Japan and southern California. Sporadically introduced in the East and occasionally recorded from, but probably no longer extant in, Lake Ontario and Georgian Bay.

BROOK TROUT—*Salvelinus fontinalis* (Mitchill). Figs. 70 and 71 and col. pl. number 5.—From the Hudson Bay and Ungava Bay drainages and Labrador coastwise to Cape Cod; southward along the Appalachians to the headwaters of the Savannah, Chattahoochee and Tennessee rivers in the Carolinas and Georgia; and in the interior to the Great Lakes basin and in a few far-northern headwaters of the upper Mississippi River system; also in southeastern Minnesota and northeastern Iowa. In the Great Lakes native to Lake Superior and tributaries, to the northern tip of the Lower Peninsula and reputedly to southern Michigan, but originally absent in most or all of the grayling waters through the northern part of the Lower Peninsula. Widely introduced in western United States and Canadian provinces, and in various other countries. In colder lakes and streams. Sea-run in the north.

Subspecies note.—In the former edition of this book, a subspecies, called *S. f. hudsonicus,* was recognized for the Cascapedia trout from Quebec, which was once introduced and locally established in the Au Sable River system of Michigan. The slenderer and more terete form of this race and its tendency to have yellow instead of bright red spots bred true in the Grayling Hatchery of the Michigan Department of Conservation for at least two generations and were recognizable in nature. Recent studies, however, have not confirmed the distinction of such a subspecies.

Lake trout—*Salvelinus namaycush* (Walbaum).—Across northern North America from northern Alaska, northern British Columbia and Canadian Arctic islands through the Labrador Peninsula to northern New England (introduced as far south as Connecticut); then westward in headwater lakes in New York of the Hudson and St. Lawrence river systems and of the Lake Ontario drainage, and through the Great Lakes basin (where confined to the main lakes and to a few cold northern lakes); also in some lakes around the northern margin of the Mississippi watershed in Wisconsin and Minnesota, and in headwaters of the Fraser and Skeena rivers of British Columbia (stocks in the lakes of the Columbia and lower Fraser river systems are now thought likely to have been introduced, and early reports from Vancouver Island are discounted). Widely introduced in lakes of the West, and elsewhere. In deep water through most of the year, but spawning in the fall on rocky reefs; entering coastal waters in the Arctic.

Two localized deepwater types of distinctive appearance have been described from within the general range of the species and are retained provisionally, pending a determination of whether or not the characters are genetically fixed.

Common lake trout—*Salvelinus namaycush namaycush* (Walbaum). Fig. 72.—From the range of the species.

Siscowet—*Salvelinus namaycush siscowet* (Agassiz).—Deeper waters of Lake Superior and probably of other Great Lakes.

Rush Lake trout—*Salvelinus namaycush huronicus* (Hubbs).—Probably confined to a very deep trough in Rush Lake, Marquette County, Michigan.

Whitefish Family—*Coregonidae**

(Figs. 73–86)

In common with trouts and with graylings, the members of the whitefish family possess both an adipose fin and a process in the angle of each pelvic fin. The whitefish family is distinguished from the salmon family by the larger scales (fewer than a hundred in the lateral line), by the smaller mouth with no teeth, or very weak ones, and by internal characters. Some authorities now reunite the coregonids with the Salmonidae, and their view is favored by the seeming intermediacy of the Asiatic genus *Brachymystax.*

This family is restricted to the northern parts of Eurasia and North America. The total range corresponds roughly with the glaciated areas. All species spawn in fresh water and all but a few are confined throughout their lives to lakes and streams.

The whitefish family constitutes one of the most characteristic elements in the fauna of the Great Lakes. There are three principal types of fishes in the group: the lake whitefish and the shallowwater ciscoes ("lake herring") and deepwater ciscoes or "chubs," and the round whitefish or "menominee." All of these and their varieties inhabit the clear, cold or cool waters of the Great Lakes proper and deeper inland lakes. The "chubs" range down to depths of more than 700 feet, but the other species generally do not occur in water deeper than 180 feet. Most are fall spawners, broadcasting their eggs on the bottom in shoal areas about Thanksgiving time. Exceptions are among the "chubs," which spawn at various times of the year; in fact, it has been said that some species of "chub" is spawning in every month except July.

* The coregonids are almost too variable to permit the construction of a satisfactory key for their identification. Reference should be made to the distributional list in this book for ranges, to the works of Koelz (1929 and 1931), and to Chart 1, p. 52, for characters.

The lake whitefish has been the most valuable food fish in the Great Lakes. It is also caught sparingly by anglers. The average size taken by commercial fishermen was about twenty-two inches and 3½ pounds. The species was formerly artificially propagated on a large scale in the region but this activity is now greatly curtailed since it has never been shown that the millions of fry planted at considerable expense over the last fifty years or more have increased or even maintained the supply. The lake whitefish is seriously depleted at present and only drastic and uniform regulation and reduction of the sea lamprey population can hope to restore the fishery.

The ciscoes and the "chubs" are not as good food and are not as valuable as the whitefish. They are, however, particularly delicious when smoked. The sizes caught are variable. The ciscoes average the larger (but not as large as whitefish). Most of the "chubs" are small species.

The round whitefish is the least valuable of the coregonids to the commercial fisherman. The usual size entering the catch is about two pounds.

1. A single flap between the nostrils (Fig. 28A); gill-rakers fewer than 20 (Fig. 29A); body subterete (subgenus *Prosopium*) 2
   Two flaps between the nostrils (Fig. 28B); gill-rakers more than 23 (Fig. 29 B, C); body usually laterally compressed (*Coregonus*) 3

2. Lateral line scales most often 56 to 66; scale rows around body usually 33 to 37; scales around caudal peduncle 18 to 20; pyloric caeca 15 to 23; vertebrae 52 to 54; total length less than 8 inches.
   PYGMY WHITEFISH—*Prosopium coulteri* (Eigenmann and Eigenmann). (Fig. 86)
   Lateral line scales most often 83 to 96; scale rows around body usually 42 to 46; scales around caudal peduncle 22 to 24; pyloric caeca 87 to 117; vertebrae 59 to 63; total length exceeding 20 inches.
   ROUND WHITEFISH—*Prosopium cylindraceum quadrilaterale* (Richardson). (Fig. 85)

3. Premaxillaries (at front of upper jaw) wider than long, retrorse in position, giving front of snout a rounded, rather blunt profile (Fig. 30A); maxillary usually contained not less than 3 times in head (but less than 3.8); lower jaw usually contained not less than 2.4 to 2.7 times in head; gill-rakers fewer than 32 (Fig. 29B).
   LAKE WHITEFISH—*Coregonus clupeaformis* (Mitchill).* (Fig. 84)
   Premaxillaries longer than wide, usually antrorse in position, giving front of snout an angular profile (Fig. 30B), never retrorse; maxillary seldom contained more than 3 times in head; gill-rakers usually more than 31 (Fig. 29C) (subgenus *Leucichthys*, Chart 1.) 4

4. Body more or less ovate, deeper forward than medially (deep-water species of Great Lakes) 5
   Body elliptical or nearly so, deepest medially 11

5. Gill-rakers usually fewer than 33.
   DEEPWATER CISCO—*Coregonus johannae* (Wagner). (Fig. 81)
   Gill-rakers usually more than 33 6

6. Small, slab-sided fishes; mandible thin, with a symphyseal knob (*Coregonus kiyi*) 7
   Large, thick fishes; mandible thick, without symphyseal knob (*Coregonus nigripinnis*) 8

* For subspecies refer to Koelz (1931) and to the distributional list in this book, but be aware of the plasticity of coregonids and of the uncertainties of subspecific identities in the genus *Coregonus*.

7. Gill-rakers usually 36 to 41.
   MICHIGAN KIYI—*Coregonus kiyi kiyi* (Koelz). (Fig. 82)
   Gill-rakers usually 43 to 46.
   ONTARIO KIYI—*Coregonus kiyi orientalis* (Koelz).

8. Fins largely black; gill-rakers usually 46 to 51 9
   Fins not largely black; gill-rakers usually 38 to 42 10

9. Scales in lateral line usually 80 to 87.
   MICHIGAN BLACKFIN CISCO—*Coregonus nigripinnis nigripinnis* (Gill). (Fig. 83)
   Scales in lateral line usually 70 to 77.
   NIPIGON BLACKFIN CISCO—*Coregonus nigripinnis regalis* (Koelz).

10. Lake Superior.
    SUPERIOR BLACKFIN CISCO—*Coregonus nigripinnis cyanopterus* (Jordan and Evermann).
    Lake Ontario (probably extinct).
    ONTARIO BLACKFIN CISCO—*Coregonus nigripinnis prognathus* Smith.

11. Gill-rakers usually 34 to 52 12
    Gill-rakers usually 51 to 59 18

12. Gill-rakers usually 43 to 52 (excepting *C. a. greeleyi* with usually 39 to 43 gill-rakers); inland lakes and shallow waters of Great Lakes.
    SHALLOWATER CISCO—*Coregonus artedii* LeSueur. (Fig. 73)
    Gill-rakers usually 34 to 43 (usually 40 to 47 in some races of the deep-water *C. hoyi*); deep waters of Great Lakes and Siskiwit Lake on Isle Royale 13

13. Small species; mandible thin and usually with a symphyseal knob.
    GREAT LAKES BLOATER—*Coregonus hoyi* (Gill). (Fig. 80)
    Larger species; mandible thicker and usually without a symphyseal knob 14

14. Body little compressed; lower jaw included and usually blackish (*Coregonus reighardi*) 15
    Body usually more compressed; lower jaw often not projecting and not blackish 16

15. Lateral line scales usually 72 to 81.
    MICHIGAN SHORTNOSE CISCO—*Coregonus reighardi reighardi* (Koelz). (Fig. 77)
    Lateral line scales usually 66 to 77.
    SUPERIOR SHORTNOSE CISCO—*Coregonus reighardi dymondi* (Koelz).

16. Jaws usually equal.
    SHORTJAW CISCO—*Coregonus zenithicus* (Jordan and Evermann). (Fig. 78)
    Lower jaw usually projecting 17

17. Lateral line scales usually 71 to 77; confined to Siskiwit Lake on Isle Royale.
    SISKIWIT LAKE CISCO—*Coregonus bartletti* (Koelz). (Fig. 76)
    Lateral line scales usually 78 to 85; lakes Michigan and Huron.
    LONGJAW CISCO—*Coregonus alpenae* (Koelz). (Fig. 79)

18. Gill-rakers usually 51 to 54; confined to Ives Lake, northern Michigan.
    IVES LAKE CISCO—*Coregonus hubbsi* (Koelz). (Fig. 74)
    Gill-rakers usually 56 to 59; Canadian lakes.
    NIPIGON TULLIBEE—*Coregonus nipigon* (Koelz). (Fig. 75)

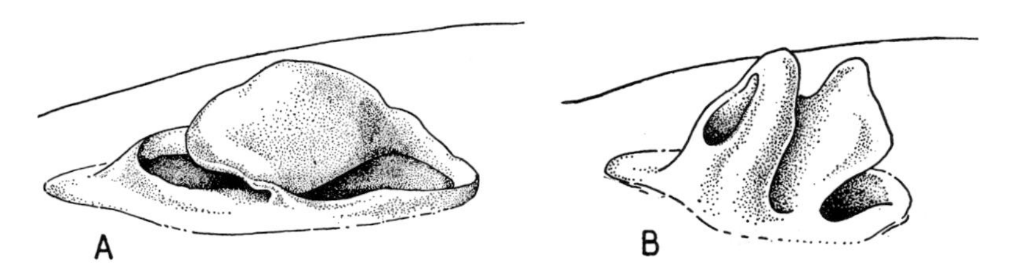

Fig. 28. A. Single flap between the nostrils (round whitefish, *Prosopium cylindraceum quadrilaterale*).

B. Two flaps between the nostrils (shallowwater cisco, *Coregonus artedii*).

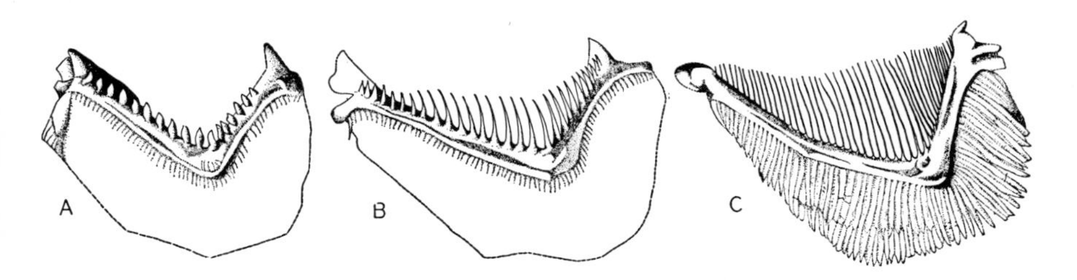

Fig. 29. Anterior gill-arches showing gill-rakers. A. Round whitefish, *Prosopium cylindraceum quadrilaterale* B. Lake whitefish, *Coregonus clupeaformis.* C. Blackfin cisco, *Coregonus nigripinnis.* (After Koelz, 1929.)

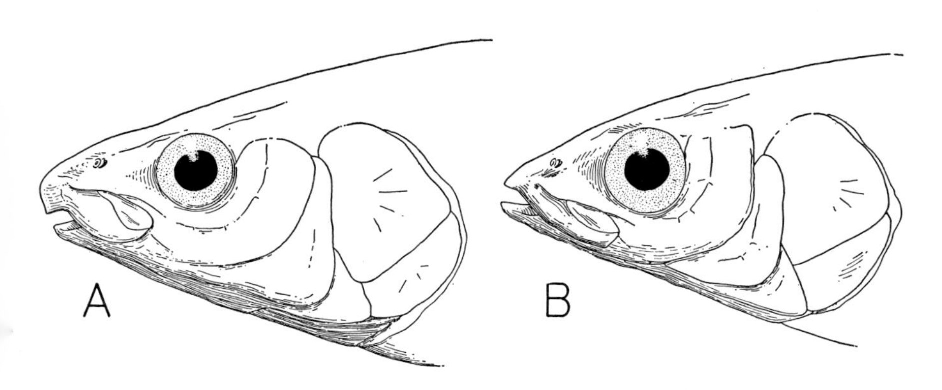

Fig. 30. Profiles of two coregonids showing difference due to position of premaxillaries. A. Premaxillaries retrorse (extending downward and backward): lake whitefish, *Coregonus clupeaformis.* B. Premaxillaries antrorse (extending downward and forward): shallowwater cisco, *Coregonus artedii.*

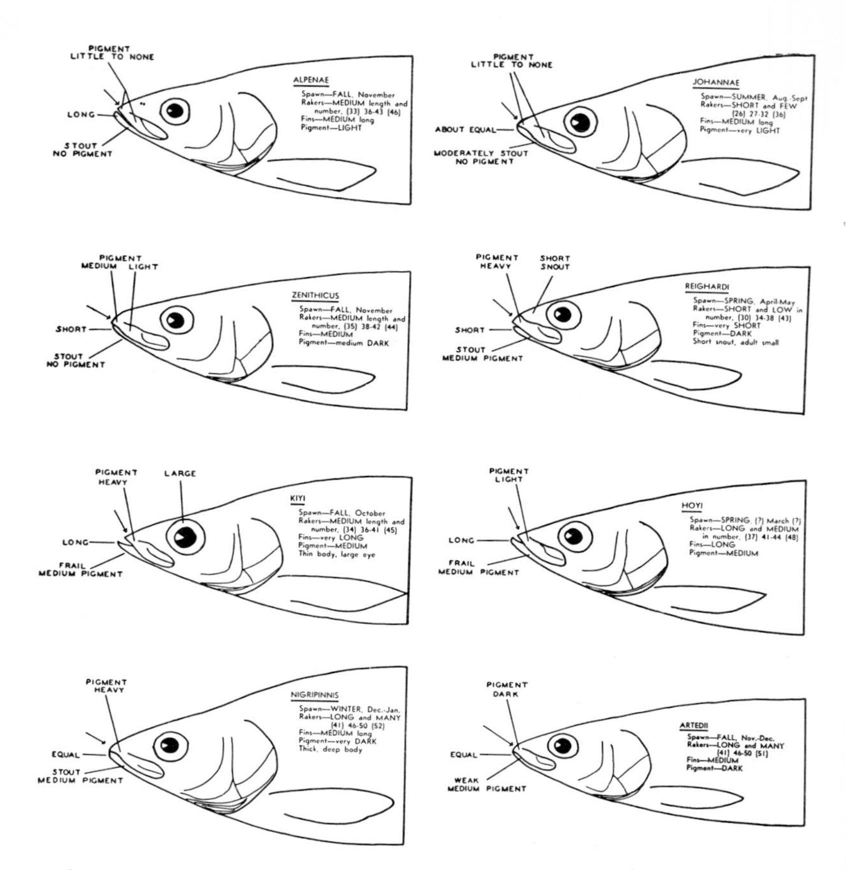

Chart 1. Summary of characteristic for separating certain species of ciscoes, genus *Coregonus*, subgenus *Leucichthys*. The arrows call attention to differences in profile of extremity of snout due to position of premaxillary bones. (Courtesy Stanford H. Smith.)

SHALLOWWATER CISCO—*Coregonus artedii* LeSueur.—In very many lakes of the Great Lakes region and the upper Mississippi River system and northward through the southern drainages of Hudson Bay; occurrence in the Mackenzie system has been doubted.

Throughout its range this species exhibits spectacular variations, many of which have been named as subspecies, particularly by Koelz (1931). Recent authors have advocated the abandonment of the subspecies, because of wide overlap in characters and because studies by Svärdson and others have shown that many of the characters are subject to environmental effects. There are indications, however, that some of the characters are in part genetic and that the gill-raker numbers are little subject to direct effects. Large plantings of eggs of the Lake Erie cisco (*albus*) in Lake Michigan are reported to have led to the appearance of fish of this type there. Provisionally, pending further research, we prefer to continue the listing of the subspecies. Difficulties in identification can always be resolved by using the binomen (*C. artedii*).

TORCH LAKE CISCO—*Coregonus artedii greeleyi* (Koelz).—Typical race in Torch and Elk lakes of the Lake Michigan basin in Michigan; also listed, in New York, from Canandaigua, Keuka, Owasco and Skaneateles lakes of the Lake Ontario watershed, Millsite Lake of the St. Lawrence drainage, Paradox Lake of the Hudson River system and Lake George of the Lake Champlain basin.

Rush Lake cisco—*Coregonus artedii huronicus* (Koelz).—Recorded only from Rush Lake, Marquette County, Michigan, and doubtfully from Lake Desor on Isle Royale and from Chautauqua Lake in New York.

Lake Anne cisco—*Coregonus artedii annensis* (Koelz).—Described from Lake Anne in the Huron Mountains, Marquette County, Michigan, in the Lake Superior basin; attributed also, with doubt, to Bernard Lake, Parry Sound Township, Ontario, in the Georgian Bay drainage.

Sargent Lake cisco—*Coregonus artedii sargenti* (Koelz).—Recorded from Sargent and Richie lakes and wrongly from Siskiwit Lake, all on Isle Royale.

Oneida Lake cisco—*Coregonus artedii hankinsoni* (Koelz).—Oneida Lake, New York, of the Lake Ontario basin.

Green Lake cisco—*Coregonus artedii birgei* (Wagner).—Green Lake, in the Lake Michigan drainage of Wisconsin.

Pine Lake cisco—*Coregonus artedii russeli* (Koelz).—Typically from Pine Lake, Marquette County, Michigan (Lake Superior drainage); also credited, with doubt, to Twelve-Mile Lake and Lake Simcoe, respectively, in the Lake Ontario and the Georgian Bay drainages of Ontario.

Whitefish Lake cisco—*Coregonus artedii atikamek* (Koelz).—Known only from Whitefish Lake, in the Lake Michigan basin of Mackinac County, Michigan, and doubtfully from Crawling Stone Lake, in the Mississippi River drainage of Vilas County, Wisconsin.

Ontario cisco—*Coregonus artedii mackayi* (Koelz).—Inland lakes in the Great Lakes basin of southern Ontario.

Great Lakes cisco—*Coregonus artedii artedii* LeSueur. Fig. 73.—In all the Great Lakes (except all or part of Lake Superior), in the upper end of the St. Lawrence River and in Lake Champlain; also in some inland lakes of Michigan (Lower Peninsula) and Wisconsin (Lake Michigan, Lake Superior and perhaps the Mississippi River watersheds); also attributed to certain New York lakes, in the Lake Ontario, St. Lawrence River and Hudson River drainages. Commonly called "lake herring" although it is not a herring (Family Clupeidae).

Seneca Lake cisco—*Coregonus artedii osmeriformis* Smith.—Seneca Lake, of the Lake Ontario basin in New York.

Clear Lake cisco—*Coregonus artedii clarensis* (Koelz).—Reported from Clear Lake and numerous other inland lakes in the Mississippi River and Lake Superior drainages of Wisconsin and in the Lake Michigan, Lake Huron and Lake Erie tributary systems of Michigan.

Nipigon cisco—*Coregonus artedii clemensi* (Koelz).—Described from Lake Nipigon, Ontario, and indicated to occur also in Lake Fanny Hoe and Mountain and Ives lakes of the Lake Superior drainage, and Gulliver Lake of the Lake Michigan basin, Michigan, and in Trout Lake of the Mississippi drainage, Wisconsin.

Hulbert Lake cisco—*Coregonus artedii lowei* (Koelz).—Types from Hulbert Lake in the Upper Peninsula of Michigan; others reported from Black Oak Lake of Vilas County, Wisconsin (both in the Lake Superior watershed).

Rawson Lake cisco—*Coregonus artedii bisselli* Bollman.—Rawson Lake and certain other inland lakes in the Lake Michigan watershed of the southern and central parts of the Lower Peninsula of Michigan.

Lake Erie cisco—*Coregonus artedii albus* LeSueur.—Typical form in Lake Erie; similar types, also referred to *albus*, in Lake Ontario, some bays of Lake Superior and in Lake Mindemoya on Manitoulin Island in Lake Huron.

Lake Superior cisco—*Coregonus artedii arcturus* (Jordan and Evermann).—Shoal and surface waters of Lake Superior, at least to the westward.

Tippecanoe Lake cisco—*Coregonus artedii sisco* (Jordan).—Tippecanoe Lake of the Wabash River system and several inland lakes in the Lake Michigan watershed of Indiana and southern Michigan; also attributed to Oconomowoc Lake, Island Lake and Lake Geneva, respectively, in the Mississippi River, Lake Superior and Lake Michigan drainages of Wisconsin.

Gogebic Lake cisco—*Coregonus artedii microcephalus* (Koelz).—Gogebic Lake of the Lake Superior basin in Michigan.

North Channel tullibee—*Coregonus artedii manitoulinus* (Jordan and Evermann).—North Channel of Georgian Bay; doubtfully also from Manitowik Lake in the Lake Superior basin of Ontario.

Twin Lake cisco—*Coregonus artedii woodi* (Koelz).—North Twin Lake and certain other lakes of the Mississippi River drainage of Wisconsin and Minnesota; also attributed to Charter and Whitestone lakes, Parry Sound County, Ontario (Georgian Bay drainage).

Lake Winnipegosis cisco—*Coregonus artedii winnipegosis* (Koelz).—Lakes Winnipegosis and Wabigoon, western Ontario, respectively tributary to Lake Manitoba and Lake Winnipeg; also identified, probably inaccurately, from Otter and Portage lakes in the Lake Superior basin of Michigan, from Van Etten Lake near Lake Huron in Michigan and from Kenogami Lake in the Timiskaming District, Ontario.

Ives Lake cisco—*Coregonus hubbsi* (Koelz). Fig. 74.—Deeper waters of Ives Lake in the Huron Mountains (Lake Superior drainage) of Michigan.

Nipigon tullibee—*Coregonus nipigon* (Koelz). Fig. 75.—Shallow waters of Lake Nipigon and Black Sturgeon Lake in the Lake Superior basin and of lakes Winnipeg and Abitibi and other, smaller lakes in northwestern Ontario, the Hudson Bay drainage and Quebec, all in Canada.

Siskiwit Lake cisco—*Coregonus bartletti* (Koelz). Fig. 76.—Known only from Siskiwit Lake, Isle Royale.

Shortnose cisco—*Coregonus reighardi* (Koelz).—This cisco, one of the so-called deepwater "chubs" of the Great Lakes commercial fisheries, has been separated into two subspecies.

Michigan shortnose cisco—*Coregonus reighardi reighardi* (Koelz). Fig. 77.—Lakes Michigan, Huron and Ontario, in water 6 to 90 fathom deep, but commonest in the shallower part of this range.

Superior shortnose cisco—*Coregonus reighardi dymondi* (Koelz).—Lakes Superior and Nipigon, in inshore waters and occasionally to depths of 50 and possibly 65 fathoms.

Shortjaw cisco—*Coregonus zenithicus* (Jordan and Evermann). Fig. 78.—Lakes Michigan, Huron, Superior and Nipigon, in 11 to 100 fathoms, chiefly in less than 30 fathoms. Also reported from lakes Winnipeg and Athabasca, from Reindeer Lake, Saskatchewan, and from the Northwest Territories of Canada.

Longjaw cisco—*Coregonus alpenae* (Koelz). Fig. 79.—Lakes Michigan and Huron, at various depths (3 to 100 fathoms), but usually in moderately deep water.

Great Lakes bloater—*Coregonus hoyi* (Gill). Fig. 80.—Lakes Ontario, Huron, Michigan, Superior and Nipigon, from marginal waters to depths of more than 100 fathoms; also reported from Eva Lake in the Rainey Lake basin (Hudson Bay watershed) of western Ontario; possibly in Lake Winnipeg.

DEEPWATER CISCO—*Coregonus johannae* (Wagner). Fig. 81.—Perhaps now extinct, but formerly in deeper waters of lakes Michigan (30 to 90 fathoms) and Huron (16 to 100 fathoms). Recently reported from deep water in the eastern part of Lake Erie.

KIYI—*Coregonus kiyi* (Koelz).—Deeper Great Lakes. One of the so-called deepwater "chubs" of the commercial fisheries, this species has been divided into two subspecies.

MICHIGAN KIYI—*Coregonus kiyi kiyi* (Koelz). Fig. 82.—Lakes Michigan, Huron and Superior, in deep water (30 to 100, generally more than 50 fathoms).

ONTARIO KIYI—*Coregonus kiyi orientalis* (Koelz).—Lake Ontario (20 to 75 fathoms).

BLACKFIN CISCO—*Coregonus nigripinnis* (Gill).—This species, best known from the Great Lakes, in the form of three subspecies, has also been reported from lakes in the southern drainage of Hudson Bay, but some doubt attends the identifications.

MICHIGAN BLACKFIN CISCO—*Coregonus nigripinnis nigripinnis* (Gill). Fig. 83.—Lakes Michigan (where rare) and Huron (where perhaps extinct), in water deeper than 30 fathoms.

SUPERIOR BLACKFIN CISCO—*Coregonus nigripinnis cyanopterus* (Jordan and Evermann).—Lake Superior (where rare), in deeper waters (15 to 100 fathoms). Also known locally as "bluefin."

NIPIGON BLACKFIN CISCO—*Coregonus nigripinnis regalis* (Koelz).—Lake Nipigon (20 to 40 fathoms) of the Lake Superior basin and Long Lake of the Hudson Bay drainage, both in Ontario, and probably in other Canadian lakes.

ONTARIO BLACKFIN CISCO—*Coregonus nigripinnis prognathus* Smith.—Lake Ontario, mostly at depths of 60 fathoms and more; now extinct or very rare. Also known locally as "bloater."

LAKE WHITEFISH—*Coregonus clupeaformis* (Mitchill).—From the headwaters of the Yukon River system in Yukon Territory and British Columbia, and from the Skeena and Fraser river systems of the latter province through the Arctic Canadian coast to Ungava Bay and Labrador; south to New England, New York, the Great Lakes basin, and the Canadian plains. Introduced in lakes from Montana to southern British Columbia. Freely entering streams and brackish water in the far north, but chiefly confined to lakes elsewhere.

The systematic treatment of the far-northern whitefishes has been much confused. Some authors suggest that *clupeaformis* may even be conspecific with *C. lavaretus* (Linnaeus) of Eurasia (also identified from Alaska). There is great variance among the populations in the Great Lakes area, where the subspecies listed below have been recognized by Koelz (1931).

GREAT LAKES WHITEFISH—*Coregonus clupeaformis clupeaformis* (Mitchill). Fig. 84.—In all the Great Lakes (excepting Lake Erie) and in Torch Lake, Michigan (and perhaps in other inland lakes from the Great Lakes basin to New England and in eastern Canada). Typically in shallow to moderate depths around the Great Lakes.

ERIE WHITEFISH—*Coregonus clupeaformis latus* Koelz.—Lake Erie; also reported from Black Bay of Lake Superior and from Walnut Lake, southern Michigan.

INLAND LAKES WHITEFISH—*Coregonus clupeaformis neohantoniensis* Prescott.—Inland lakes from Lake Athabasca to New Brunswick; south to northern New England and, in New York, to lakes of the Lake Ontario, St. Lawrence River and Hudson River systems and also Otsego Lake of the Susquehanna drainage. In the Great Lakes drainage area recorded from Lake Simcoe and supposedly the Thames River, Siskiwit Lake on Isle Royale and Stone Lake in the Lake Michigan drainage of Wisconsin.

MEDORA LAKE WHITEFISH—*Coregonus clupeaformis medorae* Koelz.—Known only from Medora Lake in Keweenaw County, Michigan (in the Lake Superior basin).

LAKE DESOR WHITEFISH—*Coregonus clupeaformis dustini* Koelz.—Described from Lake Desor on Isle Royale; erroneously attributed to Trout Lake in the Mississippi River drainage of Wisconsin.

GULLIVER LAKE WHITEFISH—*Coregonus clupeaformis gulliveri* Koelz.—Types from Gulliver Lake in the Lake Michigan drainage of the Upper Peninsula of Michigan; otherwise recorded from Chicagon Lake of the same drainage and from the Allagash River, Maine, Lake Chateaugay and other lakes in the St. Lawrence River system in northeastern New York, and Sturgeon Lake in Peterborough County, Ontario.

ROUND WHITEFISH—*Prosopium cylindraceum quadrilaterale* (Richardson). Fig. 85.—From the Bering Sea drainages of Siberia, Alaska, and British Columbia, and from the Arctic drainages from northern Alaska to Ungava Bay, southward to New Brunswick, Maine, and Connecticut, to glacial lakes in the Adirondacks (in the basins of the St. Lawrence and Hudson rivers and of Lake Ontario), to the Great Lakes region (where confined to the main lakes, including Nipigon but excluding Erie, and to the Au Sable River), and to central Canada. Entering brackish water in the far north.

The typical subspecies, *P. c. cylindraceum* (Pennant) occurs in northern Siberia west of the Bering Sea drainage. Koelz (1931) described a local form as subspecies *minor* from Chazy Lake, in the Lake Champlain drainage of New York.

PYGMY WHITEFISH—*Prosopium coulteri* (Eigenmann and Eigenmann). Fig. 86.—Isolated places in the Pacific slope, including the Columbia River basin of Washington and, in British Columbia, the same basin plus those of the Fraser, Stikine and Yukon rivers, southwestern Alaska, and throughout Lake Superior at depths between 10 and 54 fathoms. In cold waters of rivers and lakes.

GRAYLING FAMILY—*Thymallidae*

(Fig. 87)

The grayling family is separable from the Salmonidae, which it resembles most, by its enlarged, flag-like dorsal fin, with more than 15 soft-rays (fewer in salmon and trout). The Michigan-Montana grayling, which was formerly very abundant in Michigan, became extinct in the region in the late 1930's. Principal habitat of this fish was similar to that of the brook trout. Spawning was in the early spring. The eggs were laid in the shallows over sand and fine gravel. The principal foods were aquatic insects, invertebrates and sometimes small fish.

This is another Holarctic family (occurring in the northern parts of North America and Eurasia). There are only a few species, all freshwater.

Three main reasons for the extermination of this beautiful fish in Michigan have been advanced. Some say that it was caught off by anglers and by lumbermen using illegal methods of capture. Others claim that the running of logs during the spawning time of the fish in the spring gouged the eggs out of the gravel or crushed the fry. Still others hold that introduced trout were responsible as predators or competitors. All of these factors and perhaps others were no doubt involved.

SAILFIN ARCTIC GRAYLING—*Thymallus arcticus signifer* (Richardson). Fig. 87.—From the Arctic and Bering Sea drainages of northwestern Siberia through Alaska to the Stikine and Columbia rivers in British Columbia and to the western shore of Hudson Bay and to Saskatchewan. Relict populations have occurred farther south, in two widely separated regions (an early record from Lake Ontario is regarded as probably erroneous). One population, the "Michigan grayling," named *T. tricolor,* occupied the Otter and Little Carp rivers of the Lake Superior drainage of Michigan and the streams of the Lower Peninsula of Michigan from the Jordan to the Muskegon in the Lake Michigan basin and from the Cheboygan to the Rifle in the Lake Huron drainage; it persisted in the Upper Peninsula into the present century (in the Otter River until at least 1936). The last remnant there was perhaps augmented by "Montana grayling" stocked in 1914. The other relict population, the "Montana grayling," named *T. montanus*, lived, and still persists, in upper waters of the Missouri River in Montana (this "Montana grayling" was recently stocked and persisted for a time in Ford Lake, Michigan). These southern relicts (*tricolor* and *montanus*) have often been treated as distinct species or subspecies. We have found no trustworthy difference between them, and the present trend is to regard them both as inseparable from the sailfin grayling, often called "Arctic grayling."

SMELT FAMILY—*Osmeridae*

(Fig. 88)

In common with whitefishes, grayling, salmons, catfishes, and trout-perch, the smelt has on its back, in addition to an ordinary fin, an adipose fin lacking fin rays. The smelt is easily told from catfishes since the latter have many barbels and are scaleless. Along with the salmons, grayling and whitefishes, the smelt is separable from the troutperch group, since it lacks spines in its dorsal fin (the spines of the troutperch, however, are weak and require care to distinguish). The scales of the troutperch are ctenoid, whereas those of the others are cycloid. The smelt lack a gristly process in the angle of the pelvic fins (the pelvic axillary process) that members of the whitefish, salmon and grayling families possess.

This is an essentially marine family, with some representatives that run into fresh water to spawn and some kinds that are exclusively fresh water. There are fewer than two dozen species. All are Arctic or North Temperate.

In the Great Lakes region, where the smelt occurs naturally only in Lake Ontario and to the east, it has fluctuated greatly in abundance. In 1912 and subsequent years, it was introduced into the upper Great Lakes and a few inland lakes. It reached a peak of abundance in the early 1940's and then almost disappeared. Now it is apparently gaining in numbers again.

Spawning is early in the spring in tributary streams and also in shallow parts of the lakes. Principal foods appear to be small fish and invertebrates, mostly insects. The smelt attains a size of about fourteen inches and a weight of approximately one-half pound.

The smelt provides fishing in the winter through the ice for both sport and commercial purposes. During spawning runs on streams, large numbers are caught with dip nets.

Relationships of the smelt to other species of commercial importance, such as the lake trout, are as yet little understood. The smelt is known to be a predator but it is also eaten by larger fish. Whether it does more good than harm in the economy of a lake is not known.

AMERICAN SMELT—*Osmerus mordax* (Mitchill). Fig. 88.—Atlantic Coast from Labrador to vicinity of New York (*O. sergeanti* farther south); anadromous (running into streams to spawn) and landlocked. Occurring naturally in the basins of Lake Champlain, the St. Lawrence River and Lake Ontario, and now established through introduction in many other parts of the Great Lakes region, where living around the shores of the Great Lakes and in a few inland lakes.

SUCKER FAMILY—*Catostomidae*

(Figs. 89–109 and col pls. 6–11)

The suckers are fishes with only soft-rays in their fins, with cycloid scales and usually with a more or less ventral, protrusible mouth with somewhat fleshy, sucking lips. At a first glance certain small suckers may be confused with the minnow family, and *vice versa*. The two groups may be readily separated by stepping the distance from the front of the anal fin to the base of the tail fin into the distance from the front of the anal fin to the tip of the snout. If the measurement from the anal to the tail fin is contained more than two and one-half times in the distance from anal to snout, the fish is a sucker. This proportion is also true of the carp, which is a member of the minnow family, but the carp is separated from all native minnows and the suckers by its stout, spinous, first dorsal and first anal rays. If the proportion is contained less that two and one-half times, the fish is a minnow.

The suckers are very closely related to the minnows and some of them are equally as difficult to identify as some of the minnows. Difficulties in identification are largely due to the variations in proportions with size. There are nearly 100 species in North America (one penetrating south to Guatamela), one in Siberia and one or more in China. All are bottom dwellers of lakes, ponds and slow streams. Most of them ascend smaller streams to spawn in the spring, but some use the shallows of lakes. Feeding is by suction and bottom plants and animals are eaten. Growth may be very rapid in warm waters. Some of the buffalofishes may grow as large as about three feet.

Various kinds, including the common white sucker, longnose sucker, redhorses and buffaloes enter the commercial fishery in the Great Lakes and in tributary and adjacent streams. Suckers are taken commercially in gill nets, in trap nets and by dipping. They are commonly called "mullet" by commercial fishermen. They afford some angling with doughballs and worms for bait and also some spearing at night during the spawning runs in creeks. In certain waters they may become obnoxious through overpopulation and they are known to eat spawn of more valuable fish. The young are eaten in some numbers by predatory food and game fishes. Suckers are palatable, though bony.

1 Dorsal fin elongate, with more than 22 principal rays (Ictiobinae) 2

Dorsal fin short, with fewer than 20 principal rays (Catostominae) 8

2 Anterior fontanelle well developed (Fig. 31A); pharyngeal arch almost paper-thin; subopercle broadest below its middle, subtriangular (distance from eye to posteroventral angle of preopercle about equal to that from eye to upper corner of gill-cleft); intestines forming a definite helix when viewed from below (*Carpiodes*) 3

Anterior fontanelle much reduced or lacking (Fig. 31B); pharyngeal arch more or less robust (triangular in cross-section); subopercle broadest at middle, sub-semicircular (distance from eye to posteroventral angle of preopercle about three-fourths of that from eye to upper corner of gill-cleft); intestinal loops longitudinal, paralleling the sides of the body cavity and not forming a helix (*Ictiobus*). 6

3 Scales smaller, in 37 to 40 rows along body; lower lip with no trace of a median, nipple-like projection; striations on opercle weak in adults, scarcely evident in young; snout produced; tip of lower lip clearly in advance of anterior nostril; distance from tip of snout to anterior nostril equal to length of eye (much greater than eye in adults) (*Carpiodes cyprinus*) 4

Scales larger, usually in 33 to 36 rows along body; lower lip with a median nipple-like projection; opercle strongly striated in adults (weakly so in young); snout blunt, hardly produced; tip of lower lip scarcely or not at all in advance of anterior nostril; distance from tip of snout to anterior nostril less than eye (equal in large adults) 5

4 Size smaller; body deeper; eye smaller.

NORTHERN QUILLBACK CARPSUCKER—*Carpiodes cyprinus cyprinus* (LeSueur).

Size larger; body slender; eye larger.

CENTRAL QUILLBACK CARPSUCKER—*Carpiodes cyprinus hinei* Trautman. (Fig. 92.)

5. Anterior rays of dorsal fin little elongated, the longest ray not more than two-thirds length of fin. Body more slender, its depth 2.7 (in young) to 3.3 (in adults) in standard length; eye smaller; distance from tip of snout to anterior nostril contained less than 3 times in postorbital length of head.
   NORTHERN RIVER CARPSUCKER—*Carpiodes carpio carpio* (Rafinesque). (Fig. 93)

   Anterior rays of dorsal greatly elongated, the longest ray when depressed often reaching at least to posterior tip of fin (except in young). Body deep and markedly compressed, its depth 2.9 (in young) to 2.4 (in adults) in standard length; eye larger; distance from tip of snout to anterior nostril contained more than 3 times in postorbital length of head.
   HIGHFIN CARPSUCKER—*Carpiodes velifer* (Rafinesque).

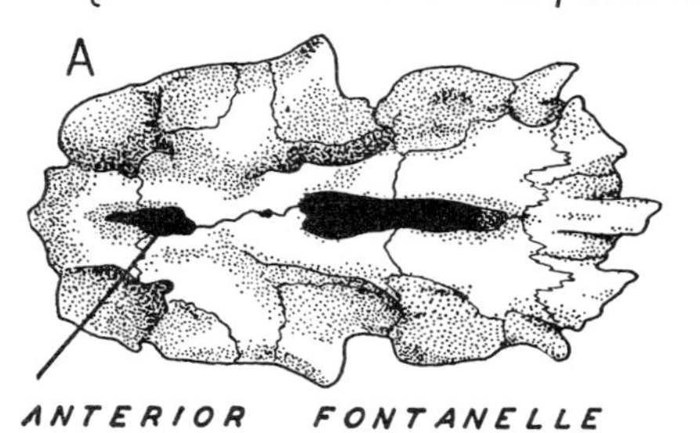

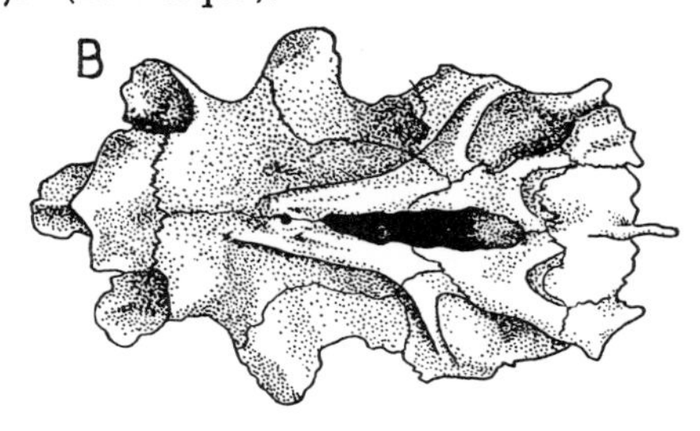

Fig. 31. Dorsal views of skulls of (A) northern river carpsucker, *Carpiodes c. carpio,* with anterior fontanelle well developed, and of (B) smallmouth buffalo, *Ictiobus bubalus,* without anterior fontanelle.

6. Mouth rather strongly oblique; lower pharyngeal arch thinner (more than twice as high as wide) and teeth weak; gill-rakers on first arch in adult nearly 100 as counted from posterior face of arch; upper jaw about as long as snout (subgenus *Megastomatobus*).
   BIGMOUTH BUFFALO—*Ictiobus cyprinellus* (Valenciennes). (Fig. 89 and col. pl. no. 6)

   Mouth horizontal or little oblique; lower pharyngeal arch heavy (about twice as wide as high) and teeth strong; gill-rakers fewer than 60 as counted from posterior face of arch; upper jaw distinctly shorter than snout (subgenus *Ictiobus*) 7

7. Body deep (depth usually about 2.5, ranging from 2.2 to 2.8 in standard length); thickness of head contained more than 5 times in standard length; distance from posterior tip of maxillary to front of mandible less than length of eye (except in large fish).
   SMALLMOUTH BUFFALO—*Ictiobus bubalus* (Rafinesque). (Fig. 91)

   Body more slender (depth usually about 3.0, ranging from 2.6 to 3.2 in standard length); thickness of head contained less than 5 times in the standard length; distance from posterior tip of maxillary to front of mandible greater than length of eye (except in young).
   BLACK BUFFALO—*Ictiobus niger* (Rafinesque). (Fig. 90)

8. Air-bladder with three chambers 9

   Air-bladder with two chambers 16

9. Premaxillaries nonprotractile; lower lip separated into two conspicuous lobes.
   HARELIP SUCKER—*Lagochila lacera* Jordan and Brayton.

   Premaxillaries protractile; lower lip with the two sides widely conjoined, not separated into two conspicuous lobes (redhorses, *Moxostoma*) 10

10 Scales around caudal peduncle 16 (7 above and 7 below lateral lines on each side 11

Scales around caudal peduncle 12 (5 above and 5 below lateral lines, rarely 13, (with 6 above) in *M. duquesnii* 12

11 Pharyngeal arch greatly enlarged and very heavy, with bases of 7 large teeth more than half length of arch (Fig. 32); middle of pupil less than halfway from snout lip to rear edge of opercle; nostril above rear margin of upper lip; scale rows between pelvic fin and lateral line 19.

COPPER REDHORSE—*Moxostoma hubbsi* Legendre. (Fig. 109)

Pharyngeal arch not greatly enlarged, nor heavy, with bases of 7 largest teeth much less than half length of arch (much as in Fig. 15); middle of pupil more than halfway from snout tip to rear edge of opercle; nostril slightly posterior to upper lip; scale rows between pelvic fin and lateral line 17.

GREATER REDHORSE—*Moxostoma valenciennesi* Jordan. (Fig. 102)

12 Lower pharyngeal arch very heavy, subtriangular in cross-section, with teeth which become enlarged downward, subcylindrical and reduced in number (Fig. 32)

RIVER REDHORSE—*Moxostoma carinatum* (Cope). (Fig. 108)

Lower pharyngeal arch only moderately heavy, distinctly narrower than high in cross-section, with teeth all markedly compressed and numerous so as to form a comb-like series (Fig. 15, p. 28) 13

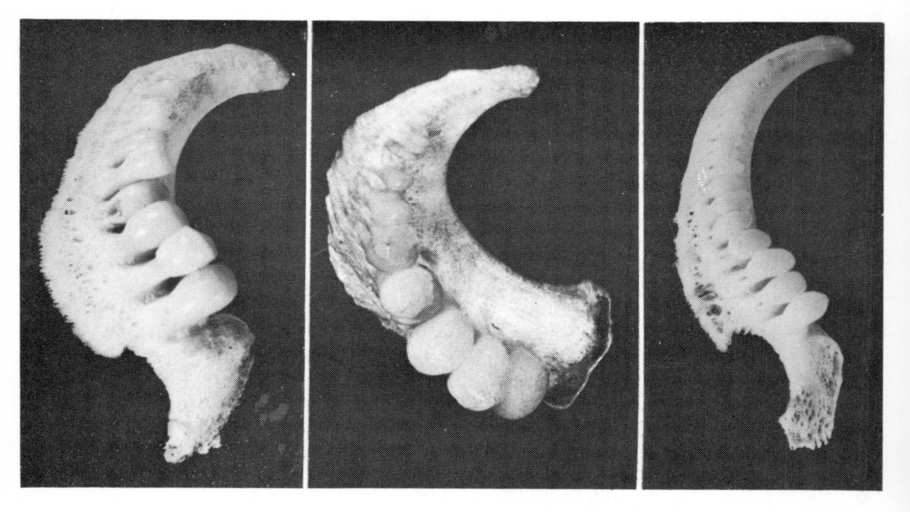

Fig. 32. Pharyngeal teeth of copper redhorse, *Moxostoma hubbsi* (left two photographs) and of the river redhorse, *Moxostoma carinatum* (right photograph). Note differences between the two species in thickness of arch and size and extent of lower teeth. (Courtesy Vianney Legendre.)

13 Body more nearly terete; caudal peduncle more slender (its least depth typically less than two-thirds its length from end of anal base); eye usually smaller, less than half snout in large young and less than two-fifths snout in small adults; scales usually smaller, typically 44 to 47 (extreme range, 42 to 49) to end of hypural; pelvic rays usually 10 (often 9 or 11).
BLACK REDHORSE—*Moxostoma duquesnii* (LeSueur). (Fig. 101)

Body less terete; caudal peduncle deeper and shorter (its least depth typically much more than two-thirds its length); eye usually larger, in young more than half length of snout, and in small adults more than two-fifths snout; scales usually larger, typically 39 to 45 (extreme range, 38 to 47) to end of hypural; pelvic rays usually 9 (often 8 or 10) 14

14 Posterior margin of lower lip nearly or quite a straight line (Fig. 33A) (in the very young an obtuse angle); mouth small; head bluntly subconical, and short (in adult contained 4.3 to 5.4 times in standard length; in young one to three inches long, about 3.5 to 3.8 times)
NORTHERN SHORTHEAD REDHORSE—*Moxostoma macrolepidotum macrolepidotum* (LeSueur). (Fig. 107 and col. pl. no. 10)

Posterior margin of lower lip an angle (Fig. 33 B, C); mouth rather large; head more squarish when seen from side, front or above, and longer (in adult contained 3.7 to 4.4 times in standard length; in young one to three inches long 3.3 to 3.7 times) 15

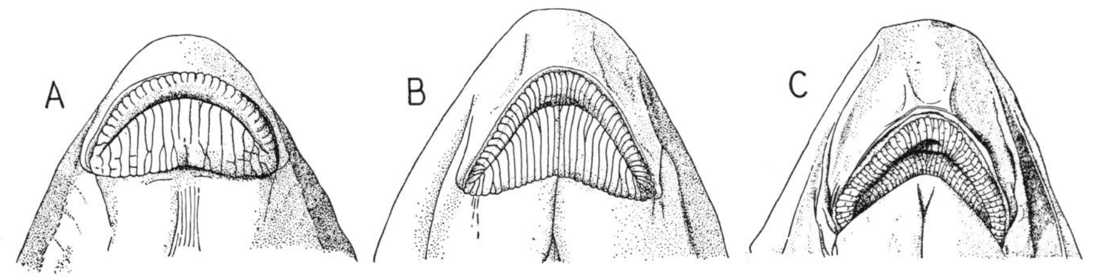

Fig. 33. Ventral views of mouths of redhorse suckers of the genus *Moxostoma*. A. Posterior margin of lower lip a straight line (northern shorthead redhorse, *M. m. macrolepidotum*. B. Posterior margin of lower lip an angle, plicae of lips (small folds) not broken into papilla-like elements (golden redhorse, *M. erythrurum*). C. Posterior margin of lower lip an angle, plicae of lips broken into papilla-like elements (silver redhorse, *M. anisurum*).

15 Plicae of lips more or less completely broken up by transverse creases into papilla-like elements (Fig. 33C); lips notably constricted; developed dorsal rays 14 to 17, usually 15 or 16; length of depressed dorsal fin more than two-thirds distance from dorsal to tip of snout; dorsal base about equal to distance from dorsal to occiput; body usually deeper (depth ordinarily contained less than 3.5 times in length in adult, in extreme variation, 3.1 to 4.1).
SILVER REDHORSE—*Moxostoma anisurum* (Rafinesque). (Figs. 105, 106)

Plicae of lips not broken by transverse creases into papilla-like elements (Fig. 33B) (except rarely to a slight degree toward angle of mouth); lips less constricted; developed dorsal rays 11 to 15, usually fewer than 15; length of depressed dorsal fin less than two-thirds distance from dorsal fin to tip of snout; dorsal base decidedly less than distance from dorsal to occiput; body slenderer as a rule (depth ordinarily contained more than 3.5 times in standard length in adult, in extreme variation, 3.35 to 4.4).
GOLDEN REDHORSE—*Moxostoma erythrurum* (Rafinesque). (Figs. 103, 104)

16. Lateral line more or less obsolescent in adult; greatest depth more than one-fifth the standard length 17

Lateral line well developed in adult; greatest depth less than one-fifth the standard length in fish larger than fingerlings 20

17. Color pattern consisting (except in the pale, obscurely mottled young) of rows of black spots, one on each scale; lateral line somewhat developed in adult; body less oblong in form, shaped as in *Moxostoma;* vertebrae about 37; mouth inferior, horizontal.

SPOTTED SUCKER—*Minytrema melanops* (Rafinesque). (Fig. 100 and col. pl. no. 8)

Color pattern consisting of two lengthwise streaks in young, more or less combined with or replaced by narrow, vertical bars in adult; lateral line wholly lacking at all ages; body more oblong; vertebrae about 34; mouth only subinferior, somewhat oblique (*Erimyzon*) 18

18. Scale rows usually 36 to 38; young with caudal fin usually reddish; large young and yearlings usually in barred-striped color phase (Fig. 97).

WESTERN LAKE CHUBSUCKER—*Erimyzon sucetta kennerlii* (Girard). (Figs. 97, 98)

Scale rows 39 to 45; young with caudal fin merely amber; large young and yearlings usually (but not always) in the juvenile striped color phase (*Erimyzon oblongus*) 19

19. Dorsal fin large, with 11 to 14 rays, usually 12; scale rows usually 41 to 43; depth in half-grown and adults contained 2.75 to 3.8 times in standard length, usually less than 3.4 times; bony interorbital width, in half-grown to adult, 2.25 to 2.65 in head.

EASTERN CREEK CHUBSUCKER—*Erimyzon oblongus oblongus* (Mitchill).

Dorsal fin smaller, with only 10 or 11 rays, usually 10; scale rows usually 39 to 41; depth in half-grown and adults contained 3.15 to 4.2 times in standard length, usually more than 3.4 times; bony interorbital width, in half-grown to adult, 2.6 to 3.0 in head.

WESTERN CREEK CHUBSUCKER—*Erimyzon oblongus claviformis* (Girard). (Fig. 99 and col. pl. no. 9)

20. Head concave above, the orbital rims broadly elevated; eye far behind middle of head; scales in fewer than 50 rows; body obliquely barred.

NORTHERN HOG SUCKER—*Hypentelium nigricans* (LeSueur). (Fig. 96 and col. pl. no. 11)

Head convex above, the orbital rim not elevated; eye near middle of head; scales in more than 50 rows; body blotched or plain (*Catostomus*) 21

21. Scales fewer than 80 in lateral line; scales subquadrate in outline and with radii more or less confined to the anterior and posterior fields (*Catostomus commersonnii*) 22

Scales more than 85 in lateral line; scales ovoid in outline and with radii more or less evenly distributed in all fields (*Catostomus catostomus*) 23

22. Scales before dorsal about 25; size generally large.

COMMON WHITE SUCKER—*Catostomus commersonnii commersonnii* (Lacépède). (Fig. 94 and col. pl. no. 7)

Scales before dorsal about 30; size smaller.

DWARF WHITE SUCKER—*Catostomus commersonnii utawana* Mather.

23 — Scales usually more than 100; size generally large.
EASTERN LONGNOSE SUCKER—*Catostomus catostomus catostomus* (Forster). (Fig. 95)

Scales about 85 to 100; size smaller.
DWARF LONGNOSE SUCKER—*Catostomus catostomus nannomyzon* Mather.

BIGMOUTH BUFFALO—*Ictiobus cyprinellus* (Valenciennes). Fig. 89 and color plate no. 6.—From Saskatchewan and the Red River of the North drainages of the Canadian plains to the Ohio Valley; south through the central part of the Mississippi Valley to Alabama, Louisiana and Texas. In the Great Lakes recorded only from western Lake Erie (where probably native). Typically an inhabitant of large rivers, oxbows, bayous and shallow lakes.

BLACK BUFFALO—*Ictiobus niger* (Rafinesque). Fig. 90.—From eastern Nebraska and Minnesota to Lake Michigan and (one report) Lake Erie, and to the Ohio Valley; south through the central parts of the Mississippi lowlands to the Gulf Coast and to Coahuila, Mexico. In sloughs and rivers, perhaps commonest in silty backwaters; in the Great Lakes common only in the marginal lakes around the southern part of Lake Michigan (where native).

SMALLMOUTH BUFFALO—*Ictiobus bubalus* (Rafinesque). Fig. 91.—From the Great Plains portion of the Hudson Bay drainage of Canada to the Ohio River system in Pennsylvania; south through the Mississippi lowlands to the Gulf Coast and the northeastern coastal drainage of Mexico. In the Great Lakes region recorded only from the St. Joseph River in southern Michigan and perhaps now extirpated in the basin. In bayous and large rivers, characteristically in the channels.

QUILLBACK CARPSUCKER—*Carpiodes cyprinus* (LeSueur).—This species has been divided into two subspecies, both of which occur in Great Lakes waters.

CENTRAL QUILLBACK CARPSUCKER—*Carpiodes cyprinus hinei* Trautman. Fig. 92.—From the southern tributaries and bays of Lake Erie in Ohio (as intergrades with *C. c. cyprinus*) through the Ohio River drainages of Ohio, Indiana, and Illinois to the vicinity of the Ohio River in West Virginia, to the larger streams of Kentucky and Tennessee, and to the Tennessee River in Alabama; also across Iowa and in eastern Missouri. Carpsuckers of this species (or species group) also occur in the Gulf drainage, western Georgia and western Florida, but require further study.

NORTHERN QUILLBACK CARPSUCKER—*Carpiodes cyprinus cyprinus* (LeSueur).—From the southern parts of Alberta and Saskatchewan and from the Red River of the North drainage of Canada and the United States, through the Great Lakes (Superior excepted) to the St. Lawrence River system (to the Ottawa River and Lake Champlain); south in the Atlantic drainage from the Susquehanna to the Roanoke; reported southward in the West to Lake of the Woods and northeastern South Dakota; material from the southern parts of Minnesota and Wisconsin and from northern Iowa has been thought to be intergrades.

HIGHFIN CARPSUCKER—*Carpiodes velifer* (Rafinesque).—Central Mississippi Valley, from Minnesota and Oklahoma to Pennsylvania and Tennessee; two records for the Great Lakes basin, for the "Root River, Michigan," more likely Root River, Wisconsin, and for Calumet River, Illinois (both in the basin of Lake Michigan). (This species was inadvertently omitted from the earlier editions of this book.)

NORTHERN RIVER CARPSUCKER—*Carpiodes carpio carpio* (Rafinesque). Fig. 93.—From the Great Plains in Montana to the Ohio River in Pennsylvania; south to the Tennessee River system in Tennessee and to the Red River between Oklahoma and Texas. In the Great Lakes drainage known from one record in the Maumee River, Ohio (introduced?). Mostly confined to large silty rivers.

WHITE SUCKER—*Catostomus commersonnii* (Lacépède).—In addition to the wide-ranging common white sucker and the dwarf derivative entered below, this species includes the weakly differentiated Great Plains white sucker, *C. c. sucklii* Girard, which ranges southward to the upper Pecos River in New Mexico and has been introduced in the Colorado River system of Colorado. In addition, several forms of the Pacific coast streams from Washington to northern Mexico are very closely related and in part perhaps conspecific.

COMMON WHITE SUCKER—*Catostomus commersonnii commersonnii* (Lacépède). Fig. 94 and color plate no. 7.—North America east of the Great Plains (and across the Great Plains if *C. c. sucklii* is the same), from the Mackenzie River in northern Canada to the Hudson Bay and Ungava Bay drainages and to the Labrador Peninsula; south on both sides of the Appalachians to Georgia and to Arkansas and northeastern Oklahoma; also in lakes of the Skeena River system in the Pacific slope of British Columbia. Widely distributed in small to large streams and in lakes, occurring most frequently in clear waters.

DWARF WHITE SUCKER—*Catostomus commersonnii utawana* Mather.—Adirondack Mountains of New York, in upland streams and lakes tributary to Lake Ontario and to the St. Lawrence and Hudson rivers.

LONGNOSE SUCKER—*Catostomus catostomus* (Forster).—From the Arctic and the Bering Sea drainages of northeastern Siberia and Alaska at least as far east as the Coppermine River, as subspecies *C. c. rostratus* Tilesius; southeastward through the ranges of the subspecies delineated below; and southward, on the east slope of the Rockies, from southern Montana to Colorado, by subspecies *C. c. griseus* Girard; represented also in the Skeena, Fraser, and upper Columbia River systems, as far south as southern Idaho (subspecies not yet worked out, except for *C. c. pocatello* Gilbert and Evermann in the Snake River system).

EASTERN LONGNOSE SUCKER—*Catostomus catostomus catostomus* (Forster). Fig. 95—From the Hudson Bay drainage of British Columbia to the Ungava Bay drainage, Labrador, and New Brunswick, south to the northern Rocky Mountains, the northern margin of the Mississippi River system in Minnesota, the southern part of the Great Lakes basin, the Youghiogheny River in Pennsylvania, the St. Lawrence River system and northeastern New England. Occurs through the Great Lakes proper, entering streams to spawn but resident in the interior only in the Lake Superior drainage. In colder lakes and streams.

DWARF LONGNOSE SUCKER—*Catostomus catostomus nannomyzon* Mather.—Known only from lakes and streams of the Adirondack and Catskill mountains of New York (Lake Ontario, Hudson and Delaware watersheds) and from the Connecticut and Merrimack river systems, and from Natashkwan, Romaine, Goynish and St. Lawrence rivers, Quebec.

WESTERN LAKE CHUBSUCKER—*Erimyzon sucetta kennerlii* (Girard). Figs. 97 and 98.—From Iowa (formerly) and from the Wisconsin River east through the southern parts of the basins of lakes Michigan, Huron, and St. Clair to Lake Erie and its tributaries including the Ontario shore, and possibly in south-central tributaries of Lake Ontario (an undescribed form?); south to southern Alabama and east-central Texas. Replaced in the southeast by *E. s. sucetta* (Lacépède). Characteristically in lakes and quieter streams.

CREEK CHUBSUCKER—*Erimyzon oblongus* (Mitchill).—This species comprises the two subspecies delineated below, plus a connectant form, *E. o. connectens* Hubbs, on the Atlantic slope of the Carolinas and Georgia.

EASTERN CREEK CHUBSUCKER—*Erimyzon oblongus oblongus* (Mitchill).—Atlantic drainage definitely from Maine, New Hampshire, and New York, southward to Virginia (at least formerly); in the Great Lakes basin confined to eastern tributaries of Lake Ontario in New York; reported in the last century from Nova Scotia, New Brunswick, and the St. Lawrence River system in Canada (it is not clear whether these reports were in error, or whether the species has disappeared in this region). Most frequent in sluggish streams.

Western creek chubsucker—*Erimyzon oblongus claviformis* (Girard). Fig. 99 and color plate no. 9.—From the southeastern corners of Wisconsin and Michigan, and the northwestern parts of Ohio and Pennsylvania, south, west of the Appalachians, to western Florida, southern Alabama, and east-central Texas. Now thought to have been reported in error from Minnesota. Chiefly a creek form of prairie regions.

Spotted sucker—*Minytrema melanops* (Rafinesque). Fig. 100 and color plate no. 8.—From Minnesota and Iowa through the drainage areas of lakes Erie and Michigan (generally rare) to the Ohio Valley (as far east as Pennsylvania); southward to the Suwannee River in northern Florida, eastern Texas, and Kansas; north on the Atlantic slope to Maryland. Predominantly in small rivers.

Black redhorse—*Moxostoma duquesnii* (LeSueur). Fig. 101.—Mississippi River and Great Lakes drainages from southern Minnesota and northern Iowa, and from the southern parts of Wisconsin, Michigan, and Ontario (where rare), and from the St. Lawrence River—Lake Champlain area; southward, west of the Appalachian divide, but chiefly in the mountains, to Alabama, Arkansas and Oklahoma. Characteristic of medium-sized, clear rivers.

Greater redhorse—*Moxostoma valenciennesi* Jordan. Fig. 102.—From the upper Mississippi River drainage in Minnesota (and questionably from Lake of the Woods), Wisconsin and Illinois through all the Great Lakes basins to the St. Lawrence River and Lake Champlain, rare or absent in the Ohio River system in Ohio; one report from the Wabash drainage in Indiana; erroneously recorded from the Tennessee River system. Usually in lakes and clear rivers. This is the species that has been known as *M. rubreques* Hubbs.

Golden redhorse—*Moxostoma erythrurum* (Rafinesque). Figs. 103 and 104.—From the Mississippi River system in Minnesota and Wisconsin and from the drainages of Lake Michigan in Wisconsin, and lakes Michigan and Huron in Michigan, and lakes St. Clair and Erie in Michigan and southern Ontario, southward, west of the mountains, to the Gulf slope of Alabama and Georgia and to Red River tributaries in Arkansas, Oklahoma and Texas. Typically in clear creeks and rivers.

Silver redhorse—*Moxostoma anisurum* (Rafinesque). Figs. 105 and 106. —From the Hudson Bay drainage of Saskatchewan and Manitoba to the St. Lawrence River basin in Quebec (to below Montreal) and to Lake Champlain; southward, west of the Appalachians, to the Tennessee River in northern Alabama and to Missouri. In the drainages of all the Great Lakes, but in that of Lake Superior known only from its western part (including Lake Nipigon). In lakes and large rivers.

Northern shorthead redhorse—*Moxostoma macrolepidotum macrolepidotum* (LeSueur). Fig. 107 and color plate no. 10.—Widespread in central and eastern Canada, eastward from the Mackenzie River region through western tributaries of James Bay, and from Saskatchewan and Montana to the entire Great Lakes—St. Lawrence watershed, including Lake Champlain; south in the east (according to Greeley) to the Hudson and Susquehanna river systems in New York, and in the west to northern Arkansas, to Kansas, and to central Wyoming; penetrating the Ohio Valley only to the Wabash River portion. Replaced through most of the Ohio River basin by *M. breviceps* (Cope) and in the Ozark region of Missouri and adjacent parts of Kansas, Oklahoma, and Arkansas by *M. m. pisolabrum* Trautman and Martin; the subspecies intergrade in northeastern Missouri. The species name *aureolum*, under which the shorthead redhorse appeared in the previous edition of this book, is not available in the genus *Moxostoma*. In muddy and clear rivers; northward commonly in lakes.

River redhorse—*Moxostoma carinatum* (Cope). Figs. 32 and 108.—From Iowa and Illinois to the Muskegon and Detroit rivers, Michigan, Lake Erie, and the St. Lawrence River in the vicinity of lakes St. Pierre and St. Louis, Quebec; south through the Ohio River system to the western parts of Pennsylvania and North Carolina and to the northern drainage of Georgia and

Alabama, and through the upper Mississippi Valley to the Ozark region of northern Arkansas and northeastern Oklahoma, where still rather common. Elsewhere generally becoming scarce. Usually in large rivers, even when young.

COPPER REDHORSE—*Moxostoma hubbsi* Legendre. Figs. 32 and 109.—St. Lawrence River from Lake St. Pierre to the mouth of Ottawa River. This is the species that was called *Megapharynx valenciennesi* (Jordan) in the last edition of this book. Recently discovered. Probably now approaching extinction.

HARELIP SUCKER—*Lagochila lacera* Jordan and Brayton.—Originally from the Maumee River, a tributary of Lake Erie in Ohio, southward to the Tennessee River system in Georgia; also in the White River of Arkansas. Not recorded for many years and possibly now exterminated.

NORTHERN HOG SUCKER—*Hypentelium nigricans* (LeSueur). Fig. 96 and color plate no. 11.—From southern Minnesota through the entire Mississippi River and Lake Michigan drainages of Wisconsin and nearby upper Michigan, the Lower Peninsula of Michigan (from the White and Au Sable rivers south), Ontario south of the Maitland River and tributaries of western Lake Ontario, and New York; south to Georgia on the east side of the Appalachians, to the Tennessee River system of Georgia and Alabama, to the Gulf slope of Mississippi, and to southwestern Arkansas and eastern Oklahoma. Replaced in the Alabama River system by *H. etowanum*. Typically found on the riffles of clear streams.

MINNOW FAMILY—*Cyprinidae*

(Figs 110–168 and col. pls. 12–19)

In common parlance, all young fish regardless of kind are called "minnows." Technically, however, minnows are fishes belonging to the family Cyprinidae, which includes daces, chubs, shiners, etc. These are all soft-rayed fishes, which, like the suckers, have toothless jaws and bear teeth in the throat only (pharyngeal teeth, Fig. 36, p. 69). Scales are cycloid.

This is probably the largest of the recognized fish families. It contains hundreds of species, which occur in almost all fresh waters in the North Temperate Zone and throughout Africa. The family is most diverse in southeastern Asia and China. Most of the North American forms are built on the same essential pattern, despite the large number of genera and species that are recognized.

Of the families occurring in the Great Lakes region the Cyprinidae is the most numerous in species and probably in individuals. Minnows occur in all types of waters—lakes, ponds, streams. Some kinds, however, have very restricted habitats. For example, the lake emerald shiner is almost entirely confined to the Great Lakes proper; certain other forms inhabit only inland lakes. More kinds occur in the southern part of the Great Lakes basin than in the northern portion.

Minnows exhibit a variety of spawning habits. Some species, such as the bluntnose minnow and relatives, make nests under boards, stones and other objects and the male guards the eggs. Others, for example the creek chub, bury the eggs in gravel and afford no other care to the eggs or young. Still others, such as the carp, broadcast the eggs upon the vegetation or bottom and, of course, desert them. Most native minnows do not reach a very large size. The maximum is about eighteen inches for the fallfish. The introduced carp, however, grows as heavy as thirty pounds or more.

Considerable economic significance may be attached to the minnows. Because of their abundance, they are very important in the food chains of all the predacious fishes. Furthermore, they are used as bait by fishermen, supporting a sizable and growing industry in bait culture in this region. Certain kinds are better suited for bait cultural work than others; the bluntnose minnow is a leader in this respect. More than one fly fisherman has thrilled to the taking of his lure by a large creek chub, river chub or common shiner.

1. Dorsal with a long base, and with more than 11 soft fin rays; dorsal and anal fins each with a strong spinous ray (Fig. 11, p. 28) (Cyprininae) 2

   Dorsal with a short base, and with fewer than 10 soft fin rays; no spinous ray in dorsal or anal fin (Leuciscinae) 3

2. Upper jaw with 2 barbels on each side; lateral line with more than 32 scales (except in "mirror" and "leather" carps).
   CARP—*Cyprinus carpio* Linnaeus. (Fig. 110 and col. pl. no. 12)

   No barbels; lateral line with fewer than 30 scales.
   GOLDFISH—*Carassius auratus* (Linnaeus). (Figs. 111, 112)

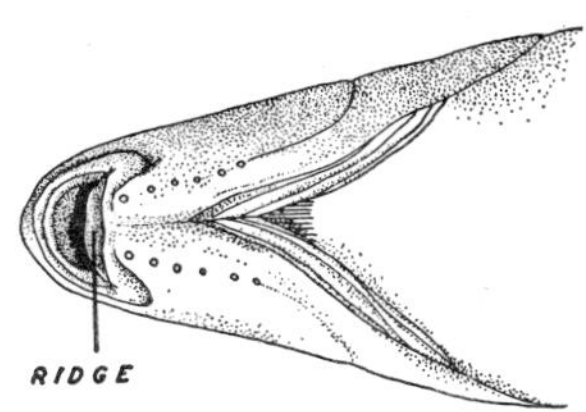

Fig. 34. Cartilaginous ridge of lower jaw prominent, separated by a definite groove from lower lip (central stoneroller, *Campostoma anomalum pullum*).

3. Cartilaginous ridge of lower jaw prominent, and separated by a definite groove from lower lip (Fig. 34); intestine spirally wound about the air-bladder (stoneroller, *Campostoma anomalum*) 4

   Cartilaginous ridge of lower jaw hardly evident, and not separated by a definite groove from lower lip; intestine not wound about air-bladder 5

4. Scales in lateral line usually 43 to 47; scale rows around body just in front of dorsal fin usually 31 to 36; width of gape usually 4.3 to 4.8 in head.
   LARGESCALE STONEROLLER—*Campostoma anomalum oligolepis* Hubbs and Greene.

   Scales in lateral line usually 48 to 51; scale rows around body just in front of dorsal fin usually 36 to 42; width of gape usually 3.7 to 4.4 in head.
   OHIO STONEROLLER—*Campostoma anomalum anomalum* (Rafinesque).

   Scales in lateral line usually 49 to 55; scales around body just in front of dorsal fin usually 39 to 46; width of gape usually 4.6 to 5.5 in head.
   CENTRAL STONEROLLER—*Campostoma anomalum pullum* (Agassiz). (Figs. 167, 168, and col. pl. no. 19)

5. Premaxillaries nonprotractile (upper lip connected with skin of snout by a frenum, a bridge of tissue across which the premaxillary groove does not pass—Fig. 35B) 6

   Premaxillaries protractile (upper lip separated from skin of snout by a deep groove continuous across the midline—Fig. 35C) 10

6. Scales in lateral line fewer than 55, and without anterior radii; lower lips more or less thickened and lobe-like; dentary bones parallel and closely approximated (Fig. 35 A, B) 7

   Scales in lateral line more than 55, and with radii all around; lower lips little modified; dentary bones divergent and well separated (as in Fig. 35C) (*Rhinichthys*) 8

7

Dentary bones devoid of fleshy tissue, looking like a tongue between the fleshy lobes of the lower lip (Fig. 35A); maxillary barbel absent.

CUTLIPS—*Exoglosum maxillingua* (LeSueur). (Fig 129)

Lower jaw and lip less modified (Fig. 35B); maxillary barbel often present.

EASTERN TONGUETIED MINNOW—*Parexoglossum laurae laurae* Hubbs. (Fig. 128)

8

Snout projecting far beyond the horizontal mouth; eyes superolateral.

GREAT LAKES LONGNOSE DACE—*Rhinichthys cataractae cataractae* (Valenciennes). (Fig. 127)

Snout scarcely projecting beyond the somewhat oblique mouth; eyes lateral 9

9

Lateral band usually more distinct than the dark speckling; caudal peduncle slender; adult males with little red on lateral band but with pectoral fins orange.

EASTERN BLACKNOSE DACE—*Rhinichthys atratulus atratulus* (Hermann).

Lateral band usually less distinct than the dark speckling (except in young); caudal peduncle deeper; adult males with lateral band red but with little color on pectoral fins.

WESTERN BLACKNOSE DACE—*Rhinichthys atratulus meleagris* Agassiz. (Fig. 126)

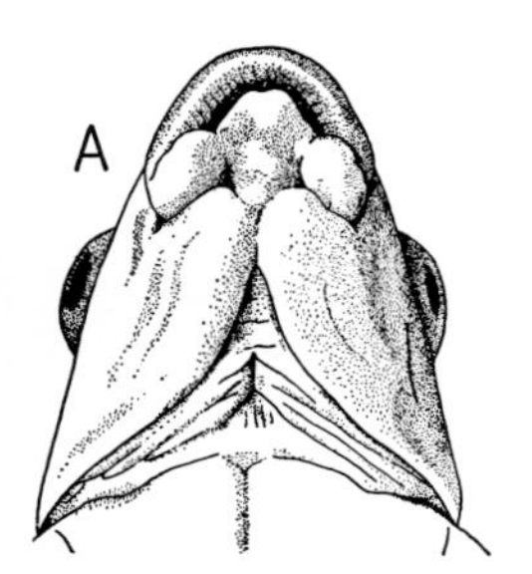

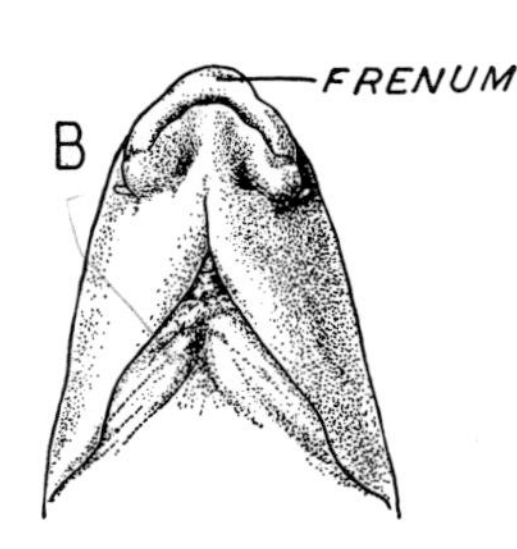

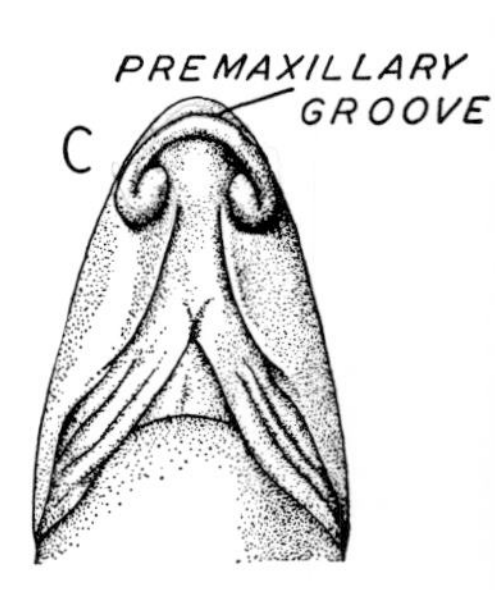

Fig. 35. Ventral views of mouths of three minnows. A. Dentary bones parallel and devoid of fleshy tissue, looking like a tongue between the fleshy lateral lobes of the lower lip (cutlips, *Exoglossum maxillingua*). B. Dentary bones parallel and lips less modified than in A (eastern tonguetied minnow, *Parexoglossum laurae laurae*). C. Dentary bones divergent, lips less modified than in A and B (suckermouth minnow, *Phenacobius mirabilis*).

10

Maxillary with a barbel (Figs. 3, 5 and 36A) (usually requiring care to observe because it is small and often hidden in the groove about the upper jaw, which should be pulled out a little in searching for the barbel; the barbel is occasionally obsolete in *Semotilus*, especially in young) 11

Maxillary without a barbel (a barbel-like swelling occurs at the end of the maxillary in breeding males of *Pimephales notatus*) 22

11
- Barbel at or near posterior end of maxillary (Fig. 3), and always slender; teeth in main row always 4–4 (*Hybopsis*) 12
- Barbel on lower edge of maxillary well in advance of posterior end (usually concealed in groove between maxillary and premaxillary, Figs. 5 and 36A, and often flap-like or obsolescent); teeth in main row typically 5–4 (Fig. 36B) (*Semotilus*) 18

12
- Scales more than 55 in lateral line (subgenus *Couesius*; northern chub, *Hybopsis plumbea*) 13
- Scales fewer than 45 in lateral line 14

13
- Length of depressed dorsal fin contained less than 1.7 times into the distance from origin of dorsal to occiput; dorsal fin typically somewhat falcate; body moderately compressed; eye larger, its diameter more than two-thirds the length of snout.
  LAKE NORTHERN CHUB–*Hybopsis plumbea plumbea* (Agassiz). (Fig. 118)
- Length of depressed dorsal fin contained more than 1.7 times into the distance from origin of dorsal to occiput; dorsal fin rounded or scarcely falcate; body terete; eye smaller, its diameter less than two-thirds the length of snout.
  CREEK NORTHERN CHUB–*Hybopsis plumbea* subspecies. (Fig. 119)

14
- Mouth somewhat oblique and inferior (upper lip scarcely overhung by snout); eye shorter than upper jaw (subgenus *Nocomis*) 15
- Mouth horizontal and strictly inferior (upper lip considerably overhung by snout); eye longer than upper jaw 16

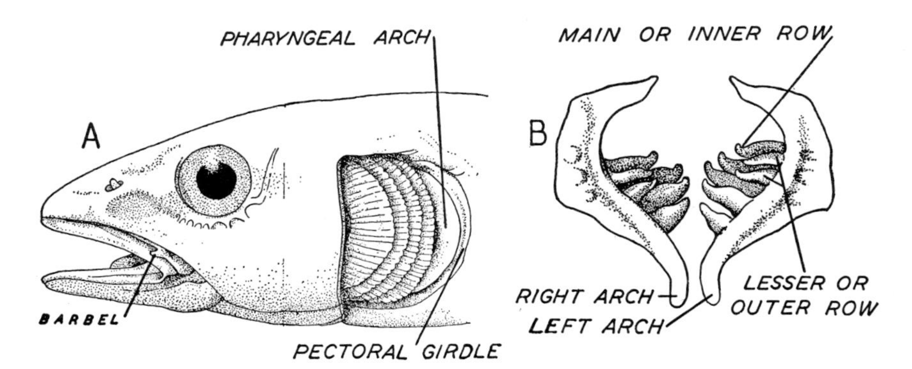

Fig. 36. Head and pharyngeal teeth of northern creek chub, *Semotilus a. atromaculatus*. A. Maxillary with a flap-like barbel well in advance of its posterior end; left pharyngeal arch, *in situ*. B. Pharyngeal arches removed and cleaned (tooth formula 2, 5–4, 2, with the teeth in main row 5–4). (After Forbes and Richardson, 1920.)

15
- Least suborbital width less than half postorbital length of head (about half in larger adults); spot at caudal base round and blackish; caudal fin red in young (in life); teeth typically 1, 4–4, 1.
  HORNYHEAD CHUB–*Hybopsis biguttata* (Kirtland). (Figs. 120, 121)
- Least suborbital width more than half postorbital length of head (except in young); spot at caudal base often indistinct, not round; caudal fin not red; teeth 4–4.
  RIVER CHUB–*Hybopsis micropogon* (Cope). (Fig. 122)

16 Body with X-shaped dark spots; teeth 4–4 (subgenus *Erimystax*).
EASTERN GRAVEL CHUB–*Hybopsis x-punctata trautmani* Hubbs and Crowe. (Fig. 125)

Body without definite spots (beware of parasite specks); teeth typically 1, 4–4, 1 (subgenus *Hybopsis*) 17

17 Color silvery, without a dark lateral band; size large, 4 to 10 inches when adult; dorsal fin inserted distinctly in advance of pelvic fins, decidedly nearer tip of snout than base of caudal fin.
SILVER CHUB–*Hybopsis storeriana* (Kirtland). (Fig. 123)

Color less silvery, with a dark lateral band; size smaller, 2 to 3 inches when adult; dorsal fin beginning over insertion of pelvic fins, usually a little nearer base of caudal than tip of snout.
BIGEYE CHUB–*Hybopsis amblops* (Rafinesque). (Fig. 124)

18 Dorsal fin beginning over pelvic base; scales large (about 45) and bright silvery (subgenus *Leucosomus*).
FALLFISH–*Semotilus corporalis* (Mitchill). (Fig. 113)

Dorsal fin beginning just behind pelvic base; scales smaller (about 50 to 75) and a little silvery 19

19 A black spot on dorsal fin near front of base (indistinct in young); mouth large (upper jaw extending at least to below front of eye); sides not mottled by specialized dark scales (subgenus *Semotilus*).
NORTHERN CREEK CHUB–*Semotilus atromaculatus atromaculatus* (Mitchill). (Figs. 36, 114–116)

No black spot on dorsal fin; mouth smaller (upper jaw not extending to below front of eye); sides mottled by specialized dark scales (subgenus *Margariscus;* pearl dace, *Semotilus margarita*) 20

20 Scales in lateral line about 50 to 60.
ALLEGHENY PEARL DACE–*Semotilus margarita margarita* (Cope).

Scales in lateral line about 65 to 75 21

21 Head more rounded and deeper (head depth usually 63 to 68 thousandths of head length); lips thicker; gape considerably curved.
NORTHERN PEARL DACE–*Semotilus margarita nachtriebi* (Cox). (Fig. 117)

Head more conic and slenderer (head depth usually 59 to 62 thousandths of head length); lips thinner; gape little curved.
HARVEY LAKE PEARL DACE–*Semotilus margarita koelzi* (Hubbs and Lagler).

22 Lateral line scales (or scale rows along side of body) more than 60 23

Lateral line scales (or scale rows along side of body) fewer than 55 27

23 Intestine short, with a single main loop, and less than twice as long as body; body with a single dusky lateral band; teeth in main row typically 5–4 (sometimes 5–5, 4–5 or 4–4) 24

Intestine elongate, with two crosswise coils in addition to the primary loop, and more than twice as long as body; body with two black lateral bands; teeth 5–5 or 5–4 (subgenus *Chrosomus*) 26

24 Lateral line incomplete; scales minute, more than 80 in lateral line, each with radii on all fields; peritoneum dark (subgenus *Pfrille*).
FINESCALE DACE– *Chrosomus neogaeus* (Cope). (Fig. 130)

Lateral line complete (except in young of *Semotilus margarita*); scales fewer than 80 in lateral line, with radii only on exposed field; peritoneum pale 25

25
- Head narrow; gape very wide, the upper jaw extending as far back as the front of the eye; nuptial tubercles small but numerous, developed on head and often on nape; snout sharp.
  REDSIDE DACE—*Clinostomus elongatus* (Kirtland). (Fig. 133)
- Head width moderate; gape much smaller, the upper jaw not extending as far back as the front of the eye; nuptial tubercles minute or undeveloped; snout blunt.
  (*Semotilus margarita*: see number 20, above in this key.)

26
- Mouth strongly oblique and more curved; length of upper jaw less than, about, or a little more than one-fourth length of head; distance from tip of snout to back of eye usually about equal to rest of head, sometimes a little longer.
  NORTHERN REDBELLY DACE—*Chrosomus eos* Cope. (Fig. 131)
- Mouth little oblique, little curved; length of upper jaw about one-fourth to considerably more than one-fourth length of head; distance from tip of snout to back of eye decidedly longer than rest of head.
  SOUTHERN REDBELLY DACE—*Chrosomus erythrogaster* Rafinesque. (Fig. 132)

27
- Abdomen behind pelvic fins with a fleshy keel over which the scales do not pass; lateral line much decurved; anal fin falcate. (golden shiner, *Notemigonus crysoleucas*) 28
- Abdomen behind pelvic fins rounded over and scaled; lateral line not greatly decurved; anal fin scarcely falcate 29

28
- Anal fin usually smaller (rays most often 12) and more rounded; body less heavy-set; fins not red in life.
  WESTERN GOLDEN SHINER—*Notemigonus crysoleucas auratus* (Rafinesque). (Fig. 135 and col. pl. no. 13).
- Anal fin usually larger (rays most often 13) and more falcate; body more heavy-set; fins red in life.
  EASTERN GOLDEN SHINER—*Notemigonus crysoleucas crysoleucas* (Mitchill).

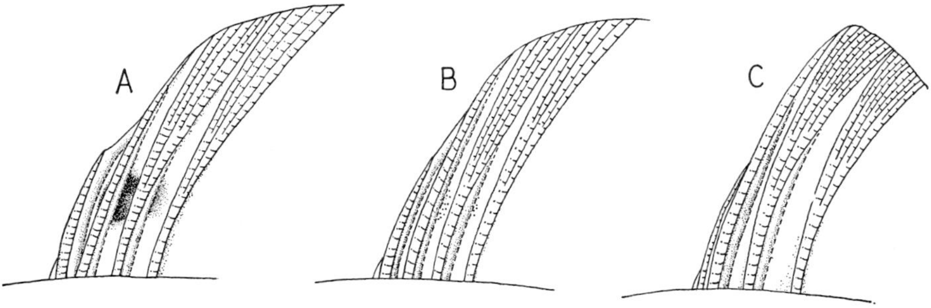

Fig. 37. Anterior portions of dorsal fins of the bluntnose minnow, *Pimephales notatus* (A, male, and B, female) and the spottail shiner, *Notropis hudsonius* (C). A and B. First obvious dorsal ray thickened and separated from first principal ray. C. First obvious dorsal ray a thin splint closely attached to first principal ray.

29 First obvious dorsal ray more or less thickened, separated by membrane from first principal ray, and with a thicker coating of dermal tissue (Fig. 37 A, B); a dark spot (faint in young and some females) at front of dorsal fin near, but not at, base; back flattish (these characters conspicuous only in adults) (*Pimephales*) 30

First obvious dorsal ray a thin splint, closely attached to first principal ray, and with a thinner covering of dermal tissue (Fig. 37C); no dark spot at front of dorsal near base (a dark spot is present at the very base in *Notropis umbratilis cyanocephalus*, Fig. 139); back little flattened 33

30 Peritoneum silvery; intestine short (S-shaped); pharyngeal teeth rather strongly hooked; nuptial tubercles on head of breeding male typically 9 (subgenus *Ceratichthys*).

NORTHERN BULLHEAD MINNOW—*Pimephales vigilax perspicuus* (Girard). (Fig. 164)

Peritoneum blackish or black; intestine somewhat elongated, with at least one short extra coil; pharyngeal teeth weakly or not at all hooked; nuptial tubercles on head of breeding male usually 16 or more 31

31 Mouth inferior and horizontal; caudal spot conspicuous; lateral line complete from pectoral arch to base of caudal fin; no nuptial tubercles on mandible of breeding males; barbel-like flap present at end of maxillary (subgenus *Hyborhynchus*).

BLUNTNOSE MINNOW—*Pimephales notatus* (Rafinesque). (Figs. 37A, B, and 165, 166)

Mouth terminal and oblique; caudal spot faint; lateral line very short (in Great Lakes subspecies); nuptial tubercles present on mandible (as well as on snout) of breeding males; barbel-like flap at end of maxillary obsolescent (subgenus *Pimephales;* fathead minnow, *Pimephales promelas*) 32

32 Lateral line scales usually 45 to 51; caudal peduncle moderately long and slender (see figures).

NORTHERN FATHEAD MINNOW—*Pimephales promelas promelas* Rafinesque. (Figs. 162, 163, and col. pl. no. 18)

Lateral line scales usually 50 to 60; caudal peduncle longer and slenderer.

HARVEY LAKE FATHEAD MINNOW—*Pimephales promelas harveyensis* Hubbs and Lagler.

33 Mouth extremely small and nearly vertical; dorsal rays typically 9; teeth in main row 5–5 or 5–4, strongly serrate.

PUGNOSE MINNOW—*Opsopoeodus emiliae* Hay. (Fig. 134)

Mouth larger and oblique to horizontal (except in *Notropis anogenus*); dorsal rays typically 8; teeth in main row 4–4 34

34 Lower lip restricted to rather prominent lateral lobes (Fig. 35C).

SUCKERMOUTH MINNOW—*Phenacobius mirabilis* (Girard). (Fig. 158 and col. pl. no. 17)

Lower lip normal, not restricted to lateral lobes 35

35 Bones of the flattened lower surface of the head with large, externally visible, cavernous chambers.

SILVERJAW MINNOW—*Ericymba buccata* Cope. (Fig. 159)

Bones of the less flattened lower surface of head without large cavernous chambers 36

36
- Anal rays 9 to 12 (rarely 8); teeth 2, 4–4, 2 37
- Anal rays usually 7 or 8, rarely 9 (usually 9 in *Notropis analostanus* and *N. l. lutrensis,* which differ from all species under items 37 to 42 in having dark pigment on the membranes between the posterior dorsal rays); teeth in outer row usually 0 or 1 (but often 2 in *Notropis heterodon, N. roseus richardsoni* and *N. hudsonius,* and typically 2 in *N. chalybaeus*) 43

37
- Origin of dorsal opposite insertion of pelvics or slightly farther forward (common shiner, *Notropis cornutus*) 38
- Origin of dorsal distinctly behind the insertion of the pelvics 39

38
- Scales before dorsal large, typically fewer than 22 (including all small and irregular scales along the midline); body averaging heavier and head deeper; dark stripes running between scale rows on back, meeting in V's posteriorly; males without dark stripe paralleling back anteriorly.
  CENTRAL COMMON SHINER—*Notropis cornutus chrysocephalus* (Rafinesque). (Fig. 140 and col. pl. no. 15)
- Scales before dorsal small, typically more than 26; body and head averaging more slender; no dark stripes between scale rows on back; males with a dark stripe paralleling back anteriorly.
  NORTHERN COMMON SHINER—*Notropis cornutus frontalis* (Agassiz). (Figs. 141, 142)

39
- Body deep, the depth typically more than head length (in adults) or equal to head length (in young); exposed portions of scales notably deeper than long; nuptial tubercles larger; breeding males with much red over entire body and fins; dorsal with prominent black spot at its extreme base, anteriorly.
  NORTHERN REDFIN SHINER—*Notropis umbratilis cyanocephalus* (Copeland). (Fig. 139)
- Body slender, the depth much less than the length of the head; exposed portions of scales not notably deeper than long; nuptial tubercles minute; breeding males with little or no red, or with red confined to the head region; dorsal fin lacking prominent black spot at its anterior base 40

40
- Snout short and blunt, its length less than two-thirds the distance from the posterior margin of the eye to the posterior end of the head; body rather sharply compressed; form of body elliptical, deepest near middle of length (emerald shiner, *Notropis atherinoides*) 41
- Snout produced and sharp, its length more than two-thirds the distance from the posterior margin of the eye to the posterior end of the head; body thicker and heavier; form of body somewhat ovate, deepest forward 42

41
- Width of body less than two-thirds height of dorsal fin (in well-preserved specimens); head in adult nearly one-fourth standard length.
  RIVER EMERALD SHINER—*Notropis atherinoides atherinoides* Rafinesque. (Col. pl. no. 14)
- Width of body more than two-thirds height of dorsal fin (in well-preserved specimens); head in adult definitely less than one-fourth standard length.
  LAKE EMERALD SHINER—*Notropis atherinoides acutus* (Lapham). (Fig. 136)

42
Stripe down middle of back jet black; origin of dorsal approximately over posterior end of the base of the pelvics; often more than three inches long.
SILVER SHINER—*Notropis photogenis* (Cope). (Fig. 137)

Stripe down middle of back merely dusky; origin of dorsal well behind base of the pelvics; seldom reaching a length of three inches.
ROSYFACE SHINER—*Notropis rubellus* (Agassiz). (Fig. 138)

43
Intestine greatly elongated (longer than twice standard length), much coiled on right side (*Hybognathus*) 44

Intestine short (much shorter than twice standard length), S-shaped .. 46

44
Dorsal fin rounded; color brassy; scales with focus less eccentric, and with about 20 radii in adult.
BRASSY MINNOW—*Hybognathus hankinsoni* Hubbs. (Fig. 160)

Dorsal fin somewhat falcate; color silvery; scales with focus very near anterior margin, and with about 10 radii in adult (*Hybognathus nuchalis*) 45

45
*Body more slab-sided; head deeper, with more evenly curved dorsal contour; snout blunter, scarcely projecting; depth of head 1.3 to 1.5 in length of head and one-fifth to one-third greater than width of body; eye in adult nearly as long as snout and about two-thirds interorbital width.
EASTERN SILVERY MINNOW—*Hybognathus nuchalis regia* Girard. (Fig. 161)

*Body wider and more turgid; head slenderer, and flatter in parietal region; snout longer and more definitely projecting; depth of head 1.5 to 1.7 in length of head, and little greater, occasionally less than width of body; eye in adult about two-thirds snout and little more than half interorbital width.
WESTERN SILVERY MINNOW—*Hybognathus nuchalis nuchalis* Agassiz.

46
Eye small, less than one-fourth length of head in adult; muzzle conical, the head subtriangular in outline; dorsal fin with a dark pigment on membranes between the posterior rays (except in young); dorsal subquadrate in outline, the last rays in adults much more than half as long as the longest 47

Eye large, more than one-fourth length of head in adult; muzzle bluntly rounded, the head not closely approaching a triangle in outline; dorsal fin without melanophores on the membranes between the posterior rays; dorsal subtriangular in outline, the last rays less than half as long as the longest 49

47
Anal rays usually 8; lateral band on posterior part of body narrow and largely below midline; membrane behind third principal dorsal ray without pigment in young and half-grown; body slender; scales in lateral line usually 39 to 42.
SPOTFIN SHINER—*Notropis spilopterus* (Cope). (Fig. 150)

Anal rays usually 9; lateral band on a posterior part of body wider and nearly median; membrane behind third dorsal ray with melanophores in young and half-grown; body deep; scales usually 34 to 38 48

* This comparison is based on an original study of material from all parts of the known range of *H. nuchalis*.

48
- Body moderately deep and thick; males not red and without enlarged nuptial tubercles above anal fin; no shoulder bar; posterior dorsal membranes blackened; teeth typically 1, 4–4, 1.
  SATINFIN SHINER—*Notropis analostanus* (Girard). (Fig. 149)
- Body deeper and thinner; males largely red, and with enlarged nuptial tubercles above anal fin; a purplish shoulder bar; posterior dorsal membranes dusky; teeth typically 4–4.
  PLAINS RED SHINER—*Notropis lutrensis lutrensis* Baird and Girard.

49
- A large, conspicuous and well-defined black spot on the base of the caudal fin; size larger, commonly more than three inches in total length (spottail shiner, *Notropis hudsonius*) 50
- No large, conspicuous and well-defined black spot on the base of the caudal fin (spot when developed not sharply set off from lateral band); size small, maximum total length about three inches 51

50
- Caudal spot smaller, less regular and more diffuse; one tooth on inner row often lacking on one or both sides.
  GREAT LAKES SPOTTAIL SHINER—*Notropis hudsonius*, subspecies. (Fig. 144)
- Caudal spot larger, more regular and more intense; full complement of teeth (2, 4–4, 2) usually developed.
  NORTHERN SPOTTAIL SHINER—*Notropis hudsonius hudsonius* (Clinton).

51
- Lateral band blackish (sometimes very indistinct in life), continued forward through eye and around muzzle; lateral line incomplete (except in *Notropis anogenus*) 52
- Lateral band dusky or obsolete, not definitely continued forward through eye and around muzzle (band rather well developed in *Notropis boops* and *N. procne*); lateral line complete 58

52
- Mouth extremely small, almost vertical; upper jaw extending only to below anterior nostril; teeth of lesser row lacking; lateral line nearly or quite complete; peritoneum blackish.
  PUGNOSE SHINER—*Notropis anogenus* Forbes. (Fig. 157)
- Mouth rather large, moderately oblique; upper jaw extending beyond anterior nostril almost to below eye; teeth of lesser row frequently developed; lateral line incomplete; peritoneum silvery 53

53
- Lateral band in surrounding muzzle confined to chin and premaxillaries; mouth more oblique, making an angle of less than 60 degrees with the vertical; teeth of lesser row (1 or 2) usually developed; blackened borders to lateral line pores not expanded into crescentic bars 54
- Lateral band in surrounding muzzle encroaching on snout above the premaxillaries, but not on the chin (chin not black); mouth less oblique, making an angle of much more than 60 degrees with the vertical; teeth of lesser row lacking; dark borders to lateral line pores expanded to form prominent, crescent-shaped, black cross-bars 56

54
- Mouth more oblique, making an angle of decidedly less than 60 degrees with the vertical; jaws equal or nearly so; snout sharp; scales of next row above lateral line on trunk with dark bars alternating with the black marks on the lateral line scales, producing a zigzag appearance.
  BLACKCHIN SHINER—*Notropis heterodon* (Cope). (Fig. 147)
- Mouth less oblique, making an angle of little less than 60 degrees with the vertical; lower jaw distinctly included; snout rather blunt; scales of next row above lateral line without definite dark bars 55

55
- Anal rays usually 7; lateral line with fewer than 10 unpored scales; nuptial tubercles in breeding males best developed on top of head; dark pigment essentially absent from inside of mouth except for a few melanophores on oral valve.
  NORTHERN WEED SHINER—*Notropis roseus richardsoni* Hubbs and Greene. (Fig. 146)
- Anal rays usually 8; lateral line with more than 10 unpored scales; nuptial tubercles in breeding males best developed on lower jaw; dark pigment conspicuous on inner borders of jaws, floor, and roof of mouth and on oral valve.
  IRONCOLOR SHINER—*Notropis chalybaeus* (Cope). (Fig. 145)

56
- Anal rays typically 8; fins smaller, length of depressed dorsal about 1.5 in distance from occiput to dorsal; dorsal usually inserted nearer base of caudal than tip of snout, and distinctly behind the pelvic insertion; predorsal scales rather irregularly arranged and 13 to 22 in number (blacknose shiner, *Notropis heterolepis*) 57
- Anal rays typically 7; fins larger, length of depressed dorsal about 1.2 in distance from occiput to dorsal; dorsal usually inserted nearer tip of snout than base of caudal, and distinctly over or before pelvic insertion; predorsal scales rather regularly arranged and usually 12 or 13 in number.
  BRIDLED SHINER—*Notropis bifrenatus* (Cope). (Fig. 156)

57
- Smaller, seldom reaching 60 mm. in standard length; head usually more than one-fourth standard length; lateral band less solidly blackened.
  NORTHERN BLACKNOSE SHINER—*Notropis heterolepis heterolepis* Eigenmann and Eigenmann. (Fig. 155)
- Larger, reaching 81 mm.; head usually about one-fourth standard length or even shorter; band more solidly blackened.
  HARVEY LAKE BLACKNOSE SHINER—*Notropis heterolepis regalis* Hubbs and Lagler.

58
- Peritoneum black; mouth rather strongly oblique.
  BIGEYE SHINER—*Notropis boops* Gilbert. (Fig. 148)
- Peritoneum silvery; mouth horizontal or nearly so 59

59
- Anal rays almost always 7 60
- Anal rays almost always 8 63

60
- Teeth usually 1 or 2, 4–4, 1 or 2; middorsal stripe not expanded at front of dorsal, but surrounding base of that fin. (Occurrence in Great Lakes basin somewhat doubtful.)
  NORTHERN RIVER SHINER—*Notropis blennius jejunus* (Forbes). (Fig. 143)
- Teeth 4–4; middorsal stripe expanded in front of dorsal and interrupted at front of dorsal base 61

61
- Lateral band somewhat developed.
  NORTHERN SWALLOWTAIL SHINER—*Notropis procne procne* (Cope). (Fig. 153)
- Lateral band scarcely developed (sand shiner, *Notropis deliciosus*) 62

62
- Length of depressed dorsal fin distinctly more than two-thirds distance from dorsal fin to occiput; scales usually 32 to 35; body more robust.
  SOUTHERN SAND SHINER—*Notropis deliciosus deliciosus* (Girard).
- Length of depressed dorsal fin about two-thirds distance from dorsal fin to occiput (or shorter); scales usually 34 to 38; body less robust.*
  NORTHEASTERN SAND SHINER—*Notropis deliciosus stramineus* (Cope). (Fig. 152 and col. pl. no. 16)

63
- Mouth small; length of upper jaw about equal to diameter of eye; exposed surface of anterior lateral line scales elevated, more than twice as high as long; teeth 4—4.*
  NORTHERN MIMIC SHINER—*Notropis volucellus volucellus* (Cope). (Fig. 154)
- Mouth large, upper jaw longer than eye; exposed surface of lateral line scales not elevated, but of usual shape; teeth 1, 4–4, 1 (bigmouth shiner, *Notropis dorsalis*) 64

64
- Form of head less extreme (approaching that of *Notropis deliciosus*).
  EASTERN BIGMOUTH SHINER—*Notropis dorsalis keimi* Fowler.
- Snout longer, mouth larger and eye smaller; head more depressed.
  CENTRAL BIGMOUTH SHINER—*Notropis dorsalis dorsalis* (Agassiz). (Fig. 151)

CARP—*Cyprinus carpio* Linnaeus. Fig. 110 and col. pl. no. 12.—Native in Asia and widely introduced elsewhere; common through most of eastern North America; in Canada north to Lake Winnipeg; including the vicinity of Nipigon Bay in Lake Superior and its tributaries; in the Great Lakes region extending northward through the Lower Peninsula of Michigan and the southern two-thirds of Wisconsin. Common in warm rivers and lakes.

GOLDFISH—*Carassius auratus* (Linnaeus). Figs. 111 and 112.—Native to eastern Asia but widely introduced elsewhere; occurring sporadically through the southern half of the Great Lakes drainage basin and in particular abundance in western Lake Erie, the Detroit River, and Lake St. Clair. Warm lakes and quiet streams, particularly in weedy situations.

FALLFISH—*Semotilus corporalis* (Mitchill). Fig. 113.—From southern tributaries of James Bay, from the Lake Superior drainage of Ontario in the vicinity of Lake Nipigon (presumably as a result of recent stream diversions from the Hudson Bay watershed), and from the Maritime Provinces, northern tributaries of the St. Lawrence River and eastern drainages of Lake Ontario, southward, east of the Appalachians, to Virginia. In streams and lakes, preferring clear ones.

NORTHERN CREEK CHUB—*Semotilus atromaculatus atromaculatus* (Mitchill). Figs. 36A (p. 69) and 114–116.—From Montana and from the Red River of the North and Lake Winnipegosis drainages to the Gaspé Peninsula in Canada, southward on both sides of the Appalachians to Georgia and other Gulf States, southwesterly to the Ozark Upland, to northeastern Texas, and to the Arkansas and upper Pecos river systems in New Mexico. Represented in eastern parts of the Gulf drainage by *S. a. thoreauianus* Jordan. In clear creeks, often very small. Known also in the Great Lakes to 7 fathoms.

PEARL DACE—*Semotilus margarita* (Cope).—We recognize three subspecies, all treated below.

* *Notropis deliciosus stramineus* and *N. volucellus volucellus* often occur together and give the beginner difficulty. Additional characters that usually distinguish the two are the distinct, narrow dark line above the main lateral band (characteristic of *N. d. stramineus*), the black pigment about the anus (in *N. v. volucellus*) and the shape of the snout as seen from above (broadly U-shaped in *N. v. volucellus*, more V-shaped in *N. d. stramineus*).

Allegheny pearl dace—*Semotilus margarita margarita* (Cope).—From Vermont and from New York (in some tributaries of lakes Erie, Ontario and Champlain and of the Allegheny, Hudson and Susquehanna rivers), south on the Atlantic slope to Virginia and in the Allegheny system of Pennsylvania. Cool, clear creeks.

Northern pearl dace—*Semotilus margarita nachtriebi* (Cox). Fig. 117. —Most of Canada south of the tundra, from the Peace River system of British Columbia to the Maritime Provinces; south to Maine, Lake Champlain and some tributaries of Lake Ontario in New York and Ontario; thence through the northern two-thirds of Michigan and most of Wisconsin to Minnesota and the Dakotas, and as rare relicts in the Sand Hills of Nebraska. Most often in cool lakes, bogs and creeks.

Harvey Lake pearl dace—*Semotilus margarita koelzi* (Hubbs and Lagler).—Confined to Harvey Lake on Isle Royale, Lake Superior. Specimens from other lakes on this island are typical of *nachtriebi* (Hatchet Lake and Lake Desor) or somewhat intermediate (Forbes Lake).

Northern chub—*Hybopsis plumbea* (Agassiz).—From the Yukon River of Alaska and Canada and the northern drainage from British Columbia to the Mackenzie River delta and Ungava Bay, southward through Labrador, Nova Scotia, and northern tributaries of the Gulf of St. Lawrence to northern New England, the Great Lakes drainage basin, and the southern tributaries of Hudson Bay; avoiding the upper Mississippi River system, except for relict populations in northeastern Iowa; on the eastern slope of the Rocky Mountains southward to south-central Wyoming (form hitherto called *dissimilis*); also, on the Pacific slope, as subspecies *H. p. greeni* (Jordan), in the Skeena, Fraser, and upper Columbia river systems.

The subspecies of *Hybopsis plumbea* are in doubt. The two forms that occur in the Keweenaw Peninsula of Michigan seem to be at least on the verge of full specific distinction, but intermediate types are reported to occur elsewhere. One of the two we previously identified as *Couesius plumbeus dissimilis,* but that name is preoccupied in *Hybopsis.*

Lake northern chub—*Hybopsis plumbea plumbea* (Agassiz). Fig. 118.—This is the common subspecies of the Great Lakes basin and, apparently, most of the northern and eastern parts of the extensive range of the species. It is particularly common in lakes, and about lakes Huron and Michigan is confined to these lakes except where it enters the coastwise streams to spawn. It is more widespread in the basins of lakes Ontario and Superior (to 12 fathoms), but has not been found in lakes Erie or St. Clair or their tributaries.

Creek northern chub—*Hybopsis plumbea,* subspecies. Fig. 119.—This is the form that lives with *H. p. plumbea* in streams of the Keweenaw Peninsula in the Lake Superior drainage of Michigan, where it appears to be an eastern outlier of the form of the eastern slopes of the Rockies hitherto generally called *dissimilis.* In Michigan, as in the west, it lives almost entirely in creeks.

Hornyhead chub—*Hybopsis biguttata* (Kirtland). Figs. 120 and 121.—From the Ozark Upland in northern Arkansas and northeastern Oklahoma and from Colorado and Wyoming to eastern North Dakota (including the Red River) and the Winnipeg River basin of Manitoba; eastward to the western half of the Lake Ontario basin in New York, the Hudson River drainage in New York and the northern part of the Ohio River system (generally avoiding the unglaciated uplands). In the Great Lakes region north to the Lake Superior and Green Bay drainages of Wisconsin and to the Straits of Mackinac; in Canada only in streams of southern Ontario in the Lake Erie and Lake Huron basins, north to Huron County. Chiefly in clear, gravelly creeks of moderate size; the younger fish commonly in weed beds.

River chub—*Hybopsis micropogon* (Cope). Fig. 122.—From the Wabash River system and the entire Lower Peninsula of Michigan to the Lake Ontario basin of Ontario and New York (west of Rochester); southward on the Atlantic slope from the Susquehanna system in New York to the James system in Virginia (records from the Passaic River in New Jersey and from near

Woods Hole in Massachusetts are considered erroneous), and on the uplands of the western Appalachian slope to the Tennessee River tributaries of Georgia and Alabama. In clear creeks and rivers (more commonly in rivers to the northward).

SILVER CHUB—*Hybopsis storeriana* (Kirtland). Fig. 123.—From the drainages of the Red River in Canada and of Lake Erie in Michigan, Ontario and Ohio, and from the south shore of Lake Ontario southward, west of the Appalachians, to the Alabama River system, to the Red River in Arkansas and to Oklahoma and Texas and to eastern Wyoming. In large, silty rivers and in lakes.

BIGEYE CHUB—*Hybopsis amblops* (Rafinesque). Fig. 124.—From the Ozark Upland of Missouri, Oklahoma and Arkansas through the southeastern half of Illinois to southeastern Michigan and to southern tributaries of the western half of Lake Ontario; thence south through the entire Ohio River system to tributaries of the Tennessee River in northern Alabama. Usually in creeks and small rivers, sandy or gravelly, clear or silty.

EASTERN GRAVEL CHUB—*Hybopsis x-punctata trautmani* Hubbs and Crowe. See Fig. 125—Ohio River basin, in Illinois, Indiana, Ohio, New York, Pennsylvania, and Kentucky (avoiding the Cumberland and Tennessee river systems); an isolated population, perhaps now extinct, in the Thames River of the Lake St. Clair drainage of Ontario (an old record from the Lake Erie drainage of northern Ohio is now doubted); presumably intergrading in Illinois with *H. x. x-punctata* Hubbs and Trautman, which occurs in scattered and in part dwindling populations in the Mississippi River drainages of southern Wisconsin, southeastern Minnesota, and Iowa, and abundantly, in the Ozark Upland of the southern half of Missouri, the northwestern half of Arkansas, northeastern Oklahoma, and southeastern Kansas.

BLACKNOSE DACE—*Rhinichthys atratulus* (Hermann).—This species comprises, in addition to the well-differentiated and possibly specifically distinct eastern and western forms delineated below, *R. a. obtusus* Agassiz of the Tennessee River system, and *R. a. simus* Garman of the Alabama River system.

EASTERN BLACKNOSE DACE—*Rhinichthys atratulus atratulus* (Hermann). —From the eastern end of the Lake Ontario basin and the St. Lawrence River drainage of Quebec, and from Nova Scotia and New Brunswick southward, east of the Appalachian Divide, to the Roanoke watershed in Virginia (known on the western slope only in the headwaters of the Youghiogheny in West Virginia). Cool creeks.

WESTERN BLACKNOSE DACE—*Rhinichthys atratulus meleagris* Agassiz. Fig. 126.—From northeastern Nebraska, Iowa, North Dakota, the drainage of Lake Winnipegosis in Manitoba, and the Lake of the Woods region through the entire Great Lakes basin (except about the east end of Lake Ontario) to the northern part of the Ohio River system. Almost everywhere abundant in small, cool tributaries of the Great Lakes.

GREAT LAKES LONGNOSE DACE—*Rhinichthys cataractae cataractae* (Valenciennes). Fig. 127.—In the Great Lakes drainage basin occurring around all the shores of the Great Lakes, and also in cool, swift streams in the northern two-thirds of Wisconsin, in Michigan northward from the Muskegon and Au Sable rivers, and in the St. Joseph River system of northwestern Indiana; down the St. Lawrence to the region of Montreal. Relationship to other subspecies in characters and distribution not yet precisely determined (the species ranges from coast to coast in northern North America and southward in the interior to Iowa, and along the mountains to North Carolina, northern Mexico and Oregon).

EASTERN TONGUETIED MINNOW—*Parexoglossum laurae laurae* Hubbs. Figs. 35B (p. 68) and 128.—Kanawha River system in West Virgina, Virginia and North Carolina, above the Falls of the Kanawha; Allegheny River tributaries of Pennsylvania and New York; and the upper part of the basin of Genesee River, a southern tributary of Lake Ontario. Represented in the Ohio River system of Ohio by an isolated population named *P. l. hubbsi* Trautman. Customarily in clear streams of moderate size.

Cutlips—*Exoglossum maxillingua* (LeSueur). Figs. 35A (p. 68) and 129. —From the systems of Lake Champlain and the upper St. Lawrence River in Quebec and Ontario and from the eastern part of the Lake Ontario watershed (including Cayuga Lake) southward, east of the Appalachians, to the Roanoke River drainage of Virginia. Ordinarily in clear, gravely creeks.

Finescale dace—*Chrosomus neogaeus* (Cope). Fig. 130.—From the Northwest Territories of Canada through the southern drainage of Hudson Bay to New Brunswick; south to Maine, New Hampshire and the Adirondack region (Lake Champlain, upper Hudson River, St. Lawrence River and Lake Ontario watersheds), and to southern Ontario, the southern parts of Michigan and Wisconsin (rare southward) and northern Minnesota; with glacial relict populations in the Black Hills of South Dakota and the Sand Hills of Nebraska. Most often in cool, boggy creeks, ponds and lakes.

Northern redbelly dace—*Chrosomus eos* Cope. Fig. 131.—From the Peace-Mackenzie system of northern British Columbia and from the southern parts of the Hudson Bay drainage of Canada east to Nova Scotia; south to Maine, New Hampshire and Lake Champlain, and, as relicts, in Massachusetts, in the Hudson system of New York, in northern New Jersey, in the Susquehanna system of Pennsylvania, in the Allegheny watershed of New York and in the Patapsco and Potomac drainages of Maryland; commonly south to southern Ontario, southern Michigan, southeastern Wisconsin, Minnesota, the Dakotas, Montana and Colorado; also, as isolated glacial relicts, in the Sand Hill region of Nebraska. In all parts of the Great Lakes basin except the Lake Erie drainage of Ohio. Typical of bog ponds and creeks.

Southern redbelly dace—*Chrosomus erythrogaster* Rafinesque. Fig. 132. —From the Mississippi drainage of Iowa and southern Minnesota through the southern half of Wisconsin (both drainages) to extreme southeastern Michigan and to the Ohio River drainages of Pennsylvania and West Virginia; southward on either side of the Mississippi lowlands to the Tennessee River system in Tennessee and northern Alabama and to the northern part of the Ozark Upland in Arkansas and Oklahoma, the Missouri River drainage in Kansas and an isolated area in the Arkansas River watershed of New Mexico. Characteristic of rather cool, clear, gravelly creeks.

Redside dace—*Clinostomus elongatus* (Kirtland). Fig. 133.—Range rather discontinuous: upper Mississippi River system in northeastern Iowa (probably extirpated) and in Minnesota and Wisconsin; Lake Michigan drainage of Wisconsin; Lake Erie tributaries in Michigan (inhabiting only a few streams), Ohio (including an old record from the lake) and New York; Lake Ontario basin in Ontario (northwestern tributaries) and New York (and very rarely in the St. Lawrence system); Mohawk River tributaries in New York; Ohio River Valley in Indiana (two old records), Ohio, Kentucky (rare), Pennsylvania and New York; Delaware River at Trenton (now extirpated?); and through the Susquehanna River system (where perhaps grading toward *C. "vandoisulus"*). The range is rapidly contracting due to deteriorating stream conditions. Typically in clear, gravelly creeks.

Pugnose minnow—*Opsopoeodus emiliae* Hay. Fig. 134.—From eastern Iowa, southern Minnesota, and southern Wisconsin to Lake St. Clair, the Detroit River and western Lake Erie; southwest in the lowlands to Texas, and southeast to extreme western Florida; replaced farther east by a form (subspecies?) with less oblique mouth. In sluggish, generally weedy waters.

Golden shiner—*Notemigonus crysoleucas* (Mitchill).—The range of the species covers that of the two subspecies listed below, plus that of the southeastern form *N. c. bosci* (Valenciennes), ranging to Florida, and *N. c. seco* (Girard), ranging to Texas.

Some authorities do not recognize them as subspecies, because the variations form clines, but the eastern golden shiners differ from the western also in being heavier-bodied and in having redder fins, and over large areas are reasonably consistent and recognizably different.

WESTERN GOLDEN SHINER—*Notemigonus crysoleucas auratus* (Rafinesque). Fig. 135 and color plate no. 13.—From the basin of Lake Winnipeg in Manitoba, and from southeastern Saskatchewan, Wyoming, and North Dakota through the entire Great Lakes region (excepting the eastern part of the Lake Ontario basin) and the southern drainage of Hudson Bay, southward to the northern half of the Ohio River system and to the Arkansas River drainage in Arkansas and Oklahoma. Generally associated with weedy lakes and the quieter parts of streams.

EASTERN GOLDEN SHINER—*Notemigonus crysoleucas crysoleucas* (Mitchill).—From Nova Scotia, Prince Edward Island, New Brunswick, Quebec, the St. Lawrence River, and the eastern part of the Lake Ontario basin (where intergrading with *N. c. auratus*), southward on the eastern slope of the Appalachians to Virginia; then intergrading southward with *N. c. bosci* (Valenciennes). Ordinarily in ponds and sluggish streams.

EMERALD SHINER—*Notropis atherinoides* Rafinesque.—The range of this species is that of the two subspecies here recognized. The desirability of separating the small-headed terete form of Lake Michigan has been questioned. The further suggestion that the southwestern *Notropis percobromus* (Cope) is a subspecies of *atherinoides* seems unwarranted, for the two forms are readily distinguishable where they occur together.

RIVER EMERALD SHINER—*Notropis atherinoides atherinoides* Rafinesque. Col. pl. no. 14.—From Lake Athabaska and the western drainage of Hudson Bay in Saskatchewan to the Trinity River system in Texas; northeast through Lake Erie and the St. Lawrence River system to tributaries of Lake St. John in Quebec and Lake Champlain; also in the Finger Lakes, Oneida Lake and connected sluggish rivers; south to the Potomac River system. Chiefly in large rivers and shallow lakes.

LAKE EMERALD SHINER—*Notropis atherinoides acutus* (Lapham). Fig. 136.—In typical form in Lake Michigan, intergrading in Lake Huron and in Lake Superior (at least about Isle Royale) with *N. a. atherinoides*. Structurally adapted to pelagic life and extremely abundant in the open waters of Lake Michigan, moving into bays, lagoons and river mouths in spring and fall. In Michigan *N. atherinoides* is native only to the Great Lakes and to marginal waters that were covered by the Postglacial Great Lakes.

SILVER SHINER—*Notropis photogenis* (Cope). Fig. 137.—From the Wabash River system of Indiana (and possibly of Illinois), the Lake Erie tributaries of southeastern Michigan and northwestern Ohio, and the Allegheny River drainages of New York and Pennsylvania, through the Ohio River basin to the Little Tennessee River in North Carolina. Frequenting flowing pools and riffles, generally of clear and swift streams.

ROSYFACE SHINER—*Notropis rubellus* (Agassiz). Fig. 138.—From the Winnipeg River basin of Manitoba and the Red River basin in North Dakota and Minnesota through the entire Mississippi River and Lake Michigan drainages of Wisconsin, and through most of Michigan and southern Ontario (north to the Green Bay drainage of upper Michigan and the St. Marys River), to the St. Lawrence system (including Lake Champlain); southward on the east slope of the Appalachians from the Hudson River to the James River, and on the west to the New River basin of Virginia, the Ohio Valley north of the Cumberland division, and to Illinois and Iowa; similar and possibly conspecific forms occur in the Tennessee River system and the Ozark Upland. Generally in clear, swift streams and in some clear lakes.

NORTHERN REDFIN SHINER—*Notropis umbratilis cyanocephalus* (Copeland). Fig. 139.—From southern Minnesota (but not including two erroneous records from the Lake of the Woods region) and from the southern two-thirds of Wisconsin and of the Lower Peninsula of Michigan, to Ontario (southern Lake Huron, Lake Erie, and Lake St. Clair drainages only), and the Lake Ontario drainage area in New York; southward to Pennsylvania, West Virginia, Kentucky, northeastern Arkansas, eastern Missouri and Iowa; intergrading toward the southwest with *N. u. umbratilis* (Girard) of the western

Gulf drainage; replaced in the clear swift creeks of the Ohio River system by *N. ardens lythrurus.* Apparently preferring the sluggish muddy streams of prairie regions.

COMMON SHINER—*Notropis cornutus* (Mitchill).—The range of the species covers, in addition to that of the two subspecies as delineated below, the Atlantic Coast drainage (from Nova Scotia southward to the James River system in Virginia), occupied by *N. c. cornutus* (Mitchill), and the Gulf drainage (from the Red River of Oklahoma southeastward), inhabited by *N. c. isolepis* Hubbs and Brown.

CENTRAL COMMON SHINER—*Notropis cornutus chrysocephalus* (Rafinesque). Fig. 140 and col. pl. no. 15.—Southern parts of the Great Lakes—St. Lawrence drainage, as far east as Lake Champlain and reputedly in certain waters of the Hudson River system (confined in Wisconsin to the southeastern corner and in Michigan to the Lower Peninsula from the Muskegon River and Saginaw Bay southward); throughout the Ohio River system to northern Alabama; also in some of the headwater tributaries of the Alabama River in Georgia; and from the Ozark region of Missouri to the Arkansas River system in Arkansas and northeastern Oklahoma. Northward in more lowland and warmer streams and lakes; southward in the clearer and cooler creeks and small rivers.

NORTHERN COMMON SHINER—*Notropis cornutus frontalis* (Agassiz). Figs. 141, 142.—From southeastern Saskatchewan and southern Manitoba to southern Quebec, southward on the west side of the Mississippi Valley to Colorado, Kansas, Iowa and extreme northern Missouri and on the east side to the Mississippi-Ohio river system in the northern parts of Illinois, Indiana and Ohio, and in Pennsylvania and West Virginia (one locality), and to northernmost New England; possibly also southward in the Appalachians to the James River watershed in Virginia (records of *N. c. cornutus* from the Lake Ontario basin and the Susquehanna and Allegheny river systems of New York were likely based on *N. c. frontalis* or on intergrades: *chrysocephalus* × *frontalis*). Preferring cool creeks and lakes.

NORTHERN RIVER SHINER—*Notropis blennius jejunus* (Forbes). Fig. 143.—From eastern Wyoming and from the southern parts of Alberta, Saskatchewan and Manitoba through North Dakota, Minnesota, Wisconsin, Illinois, Indiana and Ohio to western Pennsylvania, southward to Kentucky and northwestern Tennessee, to Missouri, to the Mississippi River in Arkansas, and to the Sabine River in Texas, and, in the far west, to northeastern Colorado. Represented in the Arkansas River system of Arkansas, Oklahoma, and Kansas and in the Missouri River system by *N. b. blennius* (Girard), with extensive intergradation in Nebraska. In the Great Lakes basin known only from Lake Winnebego, Wisconsin, where well established, and from one somewhat doubtful record from near the Lake Michigan shore of Illinois. Characteristically in the deep, wide waters of silty rivers.

SPOTTAIL SHINER—*Notropis hudsonius* (Clinton).—The range of this species covers not only that of the two rather weakly separated though distinguishable subspecies here treated, but also that of the Atlantic Coast subspecies, *N. h. amarus* (Girard), occurring from the Delaware to the Potomac basin, and *N. h. saludanus* (Jordan and Brayton), from the James to the Ocmulgee basins, Virginia to Georgia, which together may comprise a species distinct from *hudsonius.*

GREAT LAKES SPOTTAIL SHINER—*Notropis hudsonius,* subspecies. Fig. 144.—Great Lakes and tributary waters, chiefly in the main lakes and some of the inland lakes; lacking in the Lake Superior basin. This form has customarily been identified, probably incorrectly, with the true *huldsonius* of the Hudson River.

NORTHERN SPOTTAIL SHINER—*Notropis hudsonius hudsonius* (Clinton).—Mississippi River and its main tributaries from Kansas, Iowa and Illinois northward through the Red River of the North and as far as the lower Mackenzie River and the Hudson Bay region; Lake Superior and some of its tributaries; also in the Hudson and probably the Susquehanna river systems. Usually in large rivers, over terrigenous bottom.

IRONCOLOR SHINER—*Notropis chalybaeus* (Cope). Fig. 145.—From New Jersey to central Florida and to Texas on coastal lowlands; northward, in the Mississippi Valley, to Iowa (where presumably extirpated) and northern Indiana; in the Great Lakes restricted to the St. Joseph River system (of the Lake Michigan basin), wherein it replaces *N. roseus richardsoni* Hubbs and Greene.

NORTHERN WEED SHINER—*Notropis roseus richardsoni* Hubbs and Greene. Fig. 146.—From southeastern Minnesota and eastern Iowa to the Mississippi basin in Wisconsin and Illinois and the Lake Michigan drainage of Wisconsin; also in the Grand, Kalamazoo and Saginaw river systems in Michigan and the Kankakee River system in Indiana (area of intergradation with southern subspecies, which ranges from western Florida to Texas and northward probably to southeastern Indiana, not definitely determined). Characteristic of large, weedy streams and lakes.

BLACKCHIN SHINER—*Notropis heterodon* (Cope.) Fig. 147.—Glacial lake districts of eastern North Dakota, Minnesota, eastern and northern Iowa including the Okoboji lakes, Wisconsin, northern Illinois, Michigan, Indiana, Ohio, western Pennsylvania, New York, Ontario (south of Sault Ste. Marie) and Quebec (about Montreal); in the northernmost part of the Mississippi River system, the basins of all the Great Lakes, the St. Lawrence drainage, and in headwaters, in New York, of the Susquehanna and Allegheny river systems. Typically in clear and weedy lakes.

BIGEYE SHINER—*Notropis boops* Gilbert. Fig. 148.—From the Maumee River system of Ohio (only occurrence in the Great Lakes basin) and from the Illinois River tributaries in Illinois through much of the Ohio River system in Indiana, southern Ohio and western Pennsylvania; southward in this system through Kentucky to the Cumberland and Duck drainages of Tennessee; in the west probably confined to the Ozark Upland of Missouri, Arkansas, Oklahoma and Kansas (the record for southeastern Iowa has been regarded as unacceptable, and the record from the Red River between Oklahoma and Texas was presumably based on an error or on an introduction). Becoming scarce where stream habitats have deteriorated, and presumably now extinct in the Great Lakes basin. Generally in clear creeks of limestone uplands.

SATINFIN SHINER—*Notropis analostanus* (Girard). Fig. 149.—Southern tributaries of Lake Ontario and the Hudson River system, southward, east of the Appalachians, to the Cape Fear River system in North Carolina. Chiefly frequenting creeks and reservoirs.

SPOTFIN SHINER—*Notropis spilopterus* (Cope). Fig. 150.—From eastern North Dakota to Lake Champlain, including the Great Lakes (Lake Superior excepted) and St. Lawrence River tributaries through the Ottawa River to Three Rivers, Quebec; on the Atlantic slope from the Hudson and Susquehanna systems in New York to the Potomac River system; on the eastern side of the Central Basin to the Tennessee River drainage of Alabama, and on the western side to northeastern Oklahoma. Generally in medium-sized, and often silty, rivers, but also in inland lakes and along the Great Lakes shores.

PLAINS RED SHINER—*Notropis lutrensis lutrensis* Baird and Girard.—Great Plains from Wyoming and South Dakota through southern Minnesota to northern Illinois, southward to western Kentucky, and from Mississippi to Texas, New Mexico, and northeastern Mexico. Represented in parts of Texas and in northeastern Mexico as far south as the Río Pánuco system by other subspecies (and by related species). Recently established, presumably by multiplication of escaped bait minnows, in lagoons of Lake Michigan in Chicago, and through the Colorado River delta of California, Arizona, and Baja California.

BIGMOUTH SHINER—*Notropis dorsalis* (Agassiz).—This species ranges over the territories of the two subspecies here treated, and farther west (beyond a rather wide and indefinite band of intergradation), in the Platte River system of the Great Plains in Colorado and Wyoming, where *N. d. piptolepis* (Cope) occurs in adundance.

Eastern bigmouth shiner—*Notropis dorsalis keimi* Fowler.—Allegheny River system of Pennsylvania and New York, and Genesee River and Oneida Lake in the Lake Ontario drainage of New York. Chiefly in sandy streams.

Central bigmouth shiner—*Notropis dorsalis dorsalis* (Agassiz). Fig. 151.—From central Missouri, parts of Nebraska and the Red River system of North Dakota, eastward, through the watersheds of the Mississippi River in Wisconsin and Illinois and of the Wabash River in Illinois, to the Great Lakes basin, where confined to the drainages of the Fox River in Wisconsin, to waters in the Lake Superior drainage at the base of the Keweenaw Peninsula, Michigan, to the Manistique River system in the Lake Michigan drainage of upper Michigan, to the Muskegon, Grand and Kalamazoo rivers in southern Michigan, to the Black and Rocky rivers in northern Ohio, and to tributaries of Lake Erie in New York. This species and *Ericymba buccata* both prefer shallow, sandy streams and overlap widely in distribution but originally almost never occupied the same territory. In Ohio it is now being rapidly replaced by *Ericymba* (*fide* M. B. Trautman).

Sand shiner—*Notropis deliciosus* (Girard).—The range of this species covers the areas listed below for two subspecies, plus the central Great Plains south to the Arkansas River system, where *N. d. missuriensis* Cope holds forth. Through large areas these subspecies are readily distinguishable, but toward the north *N. d. deliciosus* merges into *N. d. stramineus,* so that the distinction of the subspecies becomes less trenchant.

Southern sand shiner—*Notropis deliciosus deliciosus* (Girard).—In northeastern Mexico, Texas, the Pecos River system in New Mexico, the Red River drainage of Oklahoma, the Neosho and Missouri river watersheds of Kansas and Missouri; northward to parts of Iowa and southern Minnesota, the Dakotas, Illinois and parts of Indiana (including certain lakes in the Lake Michigan drainage of Indiana). Typically in sandy streams.

Northeastern sand shiner—*Notropis delicious stramineus* (Cope). Fig. 152 and col. pl. no. 16.—From southeastern Saskatchewan and the Lake Winnipegosis drainage of Manitoba, and from Lake of the Woods, in Ontario and Minnesota, east through the Great Lakes region to southern Ontario (as far north as Lake Nipissing) and the St. Lawrence–Champlain drainage to Lake St. Pierre), south through the Ohio and Tennessee river systems to Tennessee and southwest to parts of Illinois and Iowa. Ordinarily in sandy lakes and the flowing pools and quieter riffles of streams.

Northern swallowtail shiner—*Notropis procne procne* (Cope). Fig. 153.—From the watersheds of the Delware and Susquehanna rivers of New York and a southern tributary of Lake Ontario (presumed to have been reached through a canal), southward to the James River drainage in Virginia. Sandy streams. Farther south, from the Roanoke to the Santee system, the species is represented by *N. p. longiceps* (Cope).

Northern mimic shiner—*Notropis volucellus volucellus* (Cope). Fig. 154. —From Minnesota and from Lake Winnipeg to the Hudson Bay drainage of north-central Ontario and to Lake Champlain; through the drainages of all the Great Lakes and the upper Mississippi Valley to northern Alabama and in the west at least to Iowa (relationship to other subspecies in the Ozark region not fully elucidated); also in the Roanoke and Neuse river systems of the Atlantic slope. The range of the species continues southwestward through the Ozark region and across Texas to the Guadalupe River system, in the form of several subspecies. An ecological subspecies, *N. v. wickliffi* Trautman, inhabits the larger rivers in the upper parts of the Mississippi River system. In the Great Lakes region occurring chiefly in clear and moderately weedy lakes; farther south, generally in the pools and backwaters of creeks. An aberrant form, hitherto wrongly called *N. v. buchanani* Meek, occurs in some Lake Erie tributaries, particularly in Talbot Creek, Ontario. The true *buchanani* has recently been revalidated as a distinct species.

Blacknose shiner—*Notropis heterolepis* Eigenmann and Eigenmann.—The distributional limits of the species are indicated in the account of the typical subspecies.

Northern blacknose shiner—*Notropis heterolepis heterolepis* Eigenmann and Eigenmann. Fig. 155.—Southern Canada from Saskatchewan east—ward, including part of the Hudson Bay drainage to Quebec, New Brunswick, and Nova Scotia; south to Maine, New Hampshire and the Lake Champlain basin; to head waters of the Hudson and Susquehanna systems in New York and of the Ohio basin in New York and Pennsylvania, and to Iowa, North Dakota, and formerly, to Colorado and probably western Kansas; represented southward in the central Mississippi Valley, both east and west of the Mississippi River, by another subspecies, which is characterized by a complete lateral line (this form has long been approaching extinction). Throughout the Great Lakes region. Characteristic of weedy glacial lakes and connecting streams.

Harvey Lake blacknose shiner—*Notropis heterolepis regalis* Hubbs and Lagler.—In typical form confined to Harvey Lake on Isle Royale, Lake Superior. Other lakes on this island are populated by more or less typical forms of *N. h. heterolepis*.

Bridled shiner—*Notropis bifrenatus* (Cope). Fig. 156.—Atlantic drainage from southern Maine and New Hampshire south to the Potomac River system of Virginia, extending westward through the Lake Champlain, upper St. Lawrence River, and Lake Ontario drainages (in Ontario west only to the Bay of Quinte). Quiet, weedy waters.

Pugnose shiner—*Notropis anogenus* Forbes. Fig. 157.—From the Red River basin of eastern North Dakota through the glacial lake districts of Minnesota, Wisconsin, northern Illinois, Michigan, northern Indiana and Ohio, to the Lake Ontario and St. Lawrence drainages of New York and Ontario. In the Great Lakes basin occurring scatteringly in clear and very weedy lakes throughout the watersheds of lakes Michigan, Huron, Erie and Ontario.

Suckermouth minnow—*Phenacobius mirabilis* (Girard).—Fig. 35C (p. 68), 158 and col. pl. no. 17.—From Colorado, Wyoming and South Dakota to southern Wisconsin (Mississippi drainage) and western Ohio (Ohio and Lake Erie tributaries), southward to northwestern Tennessee and to Louisiana, central Oklahoma and eastern Texas. Characteristic of clear to silty streams of prairie regions. With the modification of streams as a result of farming, this species has been spreading its range and multiplying in Indiana and Ohio.

Brassy minnow—*Hybognathus hankinsoni* Hubbs. Fig. 160.—From the Fraser and Mackenzie drainages of British Columbia through southern Canada to the upper St. Lawrence in Quebec; southward to the Lake Champlain region, the headwaters of the Hudson River in New York, the basins of all the Great Lakes except Erie, and through all of Wisconsin; southward, farther west, through the Missouri River system of Montana, to Missouri, Nebraska, and Colorado. In creeks and lakes, most often in bog waters. Occurrence in the Fraser and Peace river systems of British Columbia has probably not been due, as has been suggested, to introduction.

Silvery minnow—*Hybognathus nuchalis* Agassiz.—The range of the silvery minnow is that of the two subspecies treated below. Recent suggestions that the Great Plains form, *H. placitus* Girard, is a subspecies of *H. nuchalis*, or is merely an environmental modification, seem unsound, in view of the wide overlap of the two types, with no clear evidence of extensive intergradation, and in view of the tendency of *nuchalis* to inhabit pools and *placitus* swift shoal waters, where they occur together.

Eastern silvery minnow—*Hybognathus nuchalis regia* Girard. Fig. 161.—From the Connecticut River and Lake Champlain drainages of New England and the Ottawa and Lake Ontario watersheds of Ontario southward, east of the Appalachians, to the Altamaha River system, Georgia. Typical habitat: larger, quieter waters.

Western silvery minnow—*Hybognathus nuchalis nuchalis* Agassiz.—Mississippi River drainage system from Montana (and perhaps the Red River of the North in Manitoba) to western Pennsylvania, southward, in typical form, at least to Arkansas and northern Alabama (coastwise material from Texas to the Alabama River system in southern Alabama is not clearly referable to subspecies and probably represents intergrades). Generally avoiding the Great Lakes basin but once reported from the Detroit River and once from the mouth of the Chicago River at the head of the Chicago Drainage Canal (these reports, not verified by us, may have been based on *H. hankinsoni*). Often in rather silty rivers, but generally in pools, backwaters, oxbows or other similar habitats.

Fathead minnow—*Pimephales promelas* Rafinesque.—This species occupies not only the range of the typical subspecies, but also extends, beyond a complicated band of intergradation into northwestern Mexico, as the southern fathead minnow, *P. p. confertus* (Girard); other subspecies occur in northern Mexico.

Northern fathead minnow—*Pimephales promelas promelas* Rafinesque. Figs. 162, 163, and col. pl. no. 18—Southern Canada from the Prairie Provinces through the western and southern drainages of Hudson Bay to the St. Lawrence drainage of Quebec; south to Maine, the Champlain basin of Vermont and New York, the Hudson and Susquehanna river systems of New York, the Ohio Valley in western Pennsylvania and New York and the Cumberland system in Tennessee; and to northern Kansas (with isolated populations in northern Oklahoma) and Colorado. Generally distributed in the Great Lakes system. Northward in boggy lakes, ponds and streams; southward and westward in silty lakes and streams.

Harvey lake fathead minnow—*Pimephales promelas harveyensis* Hubbs and Lagler.—Confined to Harvey Lake on Isle Royale, Lake Superior. Other lakes on the island are inhabited *P. p. promelas.*

Northern bullhead minnow—*Pimephales vigilax perspicuus* (Girard). Fig. 164.—From the Mississippi River in southeastern Minnesota and from central Iowa and southern Wisconsin to the Ohio River and its affluents of Indiana, southern Ohio, western West Virginia and Pennsylvania (rare); southeast to the Alabama River system in Alabama and southwest to the Arkansas River system in Oklahoma and the Trinity River system in Texas (avoiding the upper part of the Red River system in Oklahoma and Texas). The species also occurs, as *P. v. vigilax* (Baird and Girard), in the upper Red River system and south of the Trinity River, as far as the Rio Grande basin in Texas and Mexico. In the Great Lakes tributaries recorded only once from Lake St. Marys in Ohio (presumably the result of release of bait minnows) and once (perhaps erroneously) from southeastern Wisconsin. Most often in large, silty streams and bayous.

Bluntnose minnow—*Pimephales notatus* (Rafinesque). Figs. 37A, B (p. 71) and 165, 166.—From southeastern Manitoba and from tributaries of the Missouri and Red rivers in North Dakota and from Lake of the Woods through the entire Great Lakes region to southern Quebec and Lake Champlain; southward on the east side of the Appalachians from the Hudson River system in New York to the James River drainage in Virginia, and on the west side to the central Gulf states (another subspecies?) and Oklahoma. Usually in clear lakes and streams over firm bottom.

Stoneroller—*Campostoma anomalum* (Rafinesque).—As a species, the stoneroller occurs not only over the ranges of the three subspecies defined below but also southward on either side of the Alleghany Mountains at higher elevations, where another subspecies lives, and over an extensive area of the Great Plains, where *C. a. plumbeum* (Girard) holds out. Recent attempts to discredit these subspecies seem most unwise. In respect to *oligolepis*, the problem is one of possible full specific distinction, as it so often occurs with *C. a. pullum* with little or possibly no evidence of interbreeding.

Largescale stoneroller—*Campostoma anomalum oligolepis* Hubbs and Greene.—Lake Michigan and Mississippi River watersheds of Wisconsin (in the Mississippi drainage chiefly near the northern boundary of the Driftless Area); eastern Iowa (two records); and the Missouri and Arkansas river tributary systems of the northern part of the Ozark Upland, in Missouri and Arkansas. Known from clear creeks and small rivers.

OHIO STONEROLLER—*Campostoma anomalum anomalum* (Rafinesque).—At moderate elevations throughout the Ohio Valley from Indiana (east of the Wabash basin) to the northern drainage of Mississippi, Alabama and Georgia; east to the Appalachian Mountains (excluding some of the higher streams where another subspecies holds forth), transgressing the drainage divide southeastward into certain Gulf tributaries of Alabama and Georgia and northeastward into some of the southern tributaries of Lake Ontario. Showing a preference for clear brooks, creeks and small rivers; essentially a riffle form.

CENTRAL STONEROLLER—*Campostoma anomalum pullum* (Agassiz). Figs. 167, 168, and col. pl. no. 19.—From the southern parts of Minnesota, Wisconsin, and Michigan to the Lake Ontario drainage of New York; the upper Susquehanna River system; some upper tributaries of the Ohio River; and from the Wabash River system westward to Iowa, and thence southwesterly, west of the Tennessee Upland and east of the Great Plains, to northeastern Mexico and parts of New Mexico. As a rule in clear, gravelly brooks and creeks.

NORTH AMERICAN CATFISH FAMILY—*Ictaluridae*

(Figs. 169–180 and col. pls. 20–22.)

The North American catfish family is composed entirely of scaleless fishes with a first spinous ray in the dorsal fin and in each pectoral fin. All members of this group possess prominent whisker-like barbels, which are sensory in function and which give these fishes a rather feline resemblance. They are further characterized by the presence of an adipose dorsal fin.

As the name implies this family is restricted to North America. It ranges from Canada to Guatemala and is confined to fresh waters. There are fewer than fifty species. Some of the South American and some of the Asiatic freshwater families of siluroid fishes contain many more species. There are also marine families of catfish.

The local catfishes may be divided into two groups, one comprising the large species and the bullheads, which provide food for man; the other composed of the little madtoms, which carry venom glands at the base of their pectoral fin spines. The madtoms can inflict a very painful wound.

With the exception of the channel catfish, the stonecat and certain of the madtoms, our catfishes in general are inhabitants of quiet and of slowly moving waters. All are spring spawners and the females, of the bullheads at least, guard their young, convoying them about for a time after hatching. Very little is known about the rate of growth of bullheads. The flathead or mud catfish attains a weight in excess of 100 pounds and the channel cat may exceed 40 pounds. The catfishes are omnivorius and feed to a large extent upon the common fish-food invertebrates, including aquatic insects and their larvae.

All of the larger catfishes and bullheads are very fine food for humans. The flesh is white and firm. It is not necessary to skin the smaller catfishes before cooking them; skinning is a difficult task for the novice and has discouraged the use of catfish as food. Bullheads are taken readily on worms, and channel and flathead catfishes are taken on set lines. Channel catfish are also caught by floating bait downstream in fairly strong currents. Suitable bait is a strip of fish flesh cut from the belly of another fish. Favored fishing spots for channel cats are downstream from power dams, where the water is fairly rapid.

1. Adipose fin with its posterior margin free, not adnate to back nor connected with caudal fin (Figs. 169-173) 2

   Adipose fin with posterior margin adnate to the back, and separated from the fleshy upper margin of the caudal fin by not more than an incomplete notch (Figs. 174-180) 7

2. Premaxillary band of teeth with backward lateral extensions (Fig. 38); anal rays 12 to 16 (rudiments included); adipose dorsal unusually large; lower jaw strongly projecting; head greatly flattened anteriorly.

   FLATHEAD CATFISH—*Pylodictis olivaris* (Rafinesque). (Fig. 173)

   Premaxillary band of teeth without backward lateral extensions; anal rays 17 to 35 (rudiments included); adipose dorsal of moderate size; lower jaw weakly or not at all projecting; head less flattened ... 3

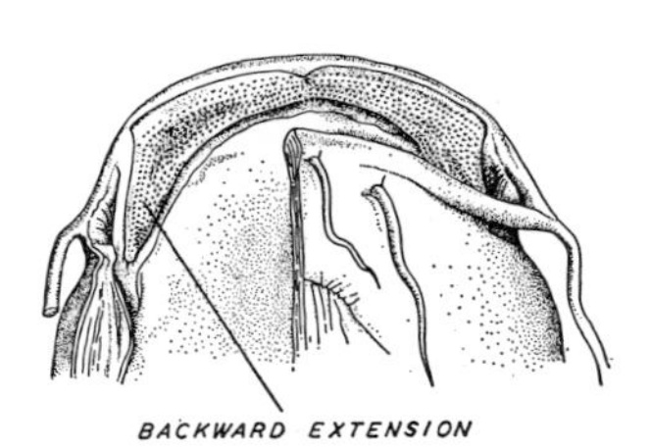

Fig. 38. Premaxillary band of teeth with backward lateral extensions, viewed from below (flathead catfish, *Pylodictis olivaris*).

3. Caudal fin forked; processes of the supraoccipital bone extended backward to form a bony ridge to, or nearly to, the origin of the dorsal fin; size larger (subgenus *Ictalurus*) 4

   Caudal fin rounded or only weakly emarginate; process of the supraoccipital bone not extended far backward, hence no bony ridge from head to dorsal fin; size smaller (subgenus *Ameiurus*) 5

4. Sides with scattered dark spots (usually); unbroken bony supraoccipital ridge from head to dorsal fin origin; lobes of tail pointed; head relatively narrow; anal rays (rudiments included) averaging about 27.

   NORTHERN CHANNEL CATFISH—*Ictalurus punctatus punctatus* (Rafinesque). (Fig. 169 and col. pl. no. 20)

   Sides plain-colored; bony ridge from head toward dorsal fin not quite complete; lobes of tail fin more or less rounded; head relatively broad; anal rays (rudiments included) averaging about 23.

   WHITE CATFISH—*Ictalurus catus* (Linnaeus)

5. Anal rays 24 to 27 (including all rudiments); caudal fin rounded; chin barbels whitish.

   NORTHERN YELLOW BULLHEAD—*Ictalurus natalis natalis* (LeSueur). (Fig. 172)

   Anal rays 17 to 24 (including all rudiments); caudal fin evidently emarginate; chin barbels gray to black 6

6. Anal fin rays (rudiments included) 21 to 24, most often 22 or 23; pectoral spine with strong barbs on posterior edge (appreciable except in old adults by the following test: grasp the spine in the plane of the fin tightly between thumb and forefinger and pull outward; if the grasp holds, the fish is *Ictalurus nebulosus*); fins more nearly unicolor, the membranes seldom jet black; lower sides often more or less mottled; no light bar at caudal base; belly in adults whitish.

   NORTHERN BROWN BULLHEAD—*Ictalurus nebulosus nebulosus* (LeSueur). (Fig. 171 and col. pl. no. 22)

   Anal fin rays (rudiments included) 17 to 21; pectoral spine without strong, definite barbs on posterior edge; fins with jet-black membranes; sides not mottled; adults with a whitish bar at caudal base; belly in adults yellowish.

   NORTHERN BLACK BULLHEAD—*Ictalurus melas melas* (Rafinesque). (Fig. 170 and col. pl. no. 21)

7. Premaxillary band of teeth with backward lateral extensions (as in Fig. 38); size moderate, not exceeding about 12 inches in total length.

   STONECAT—*Noturus flavus* Rafinesque. (Fig. 174)

   Premaxillary band of teeth without backward lateral extensions; size usually small, not exceeding about 6 inches in total length (madtoms, *Schilbeodes*) 8

8. Color of body uniform or nearly so; no strong barbs on posterior edge of pectoral spine 9

   Color of body strongly mottled; pectoral spine very strongly barbed on posterior edge 11

9. Vertical fins not dark edged; a conspicuous axial dark streak along side of body; distance from tip of caudal to notch between adipose and caudal contained once or less in distance from notch to posterior base of dorsal; procurrent caudal rays greatly developed.

   TADPOLE MADTOM—*Schilbeodes gyrinus* (Mitchill). (Fig. 175)

   Vertical fins edged with dusky or black; no axial dark streak; distance from tip of caudal to notch between adipose and caudal contained decidely more than once in distance from notch to posterior base of dorsal; procurrent caudal rays little developed 10

10. Anterior back uniform; adults usually more than 100 mm. in standard length; caudal fin with definitely evident posterior angles; head 3.6 to 3.9 times in standard length; predorsal length 2.7 to 30. times in standard length.

   COMMON EASTERN MADTOM—*Schilbeodes insignis insignis* (Richardson). (Fig. 177)

   A faint dark blotch about front of dorsal fin; adults seldom as long as 100 mm.; caudal fin more rounded; head 3.8 to 4.3 times in standard length; predorsal length 2.9 to 3.3 times in standard length. (Occurrence in Great Lakes basin doubtful.)

   SLENDER MADTOM—*Schilbeodes exilis* (Nelson). (Fig. 178)

11. Light blotch behind occiput extending to origin of dorsal fin; dark blotch on adipose fin not extending to edge of fin; distance from notch between adipose and caudal fins to end of caudal fin about 1.8 into distance from the notch to dorsal origin.

   FURIOUS MADTOM—*Schilbeodes furiosus* (Jordan and Meek). (Fig. 180)

   Light blotch behind occiput not extending to origin of dorsal fin; dark blotch on adipose extending to edge of fin; distance from notch between adipose and caudal fins to end of caudal about 1.3 into distance from the notch to dorsal origin.

   BRINDLED MADTOM—*Schilbeodes miurus* (Jordan). (Fig. 179)

NORTHERN CHANNEL CATFISH—*Ictalurus punctatus punctatus* (Rafinesque). Fig. 169 and color plate no. 20.—From the Prairie Provinces of Canada, north to 54° N. Lat., and from the southern part of the Hudson Bay drainage southward through the Great Lakes–St. Lawrence basin to the Ottawa River and through most of the Mississippi Valley (supposedly native as far west as eastern Colorado); thence to Florida and northeastern Mexico (where the subspecies are not finally determined); also in the Potomac River (possibly as a canal immigrant); widely introduced elsewhere. Represented over much of Texas and northern Mexico, south to the Río Pánuco system, by a variety of forms, of which several at least appear to be subspecies of *I. punctatus*. The far-northern form, earlier called *lacustris*, does not now appear sufficiently well set off to be recongnized, but the whole problem is in great need of critical analysis. Typically in larger rivers and in lowland lakes.

WHITE CATFISH—*Ictalurus catus* (Linnaeus).—Atlantic coastal streams from New Jersey to Lake Okeechobee in Florida, westward in tributaries of the Gulf of Mexico to western Florida. Introduced in California. Most common in slow, large rivers. In the Great Lakes recently stocked in large numbers (and established?) in Lake Erie.

NORTHERN BLACK BULLHEAD—*Ictalurus melas melas* (Rafinesque). Fig. 170 and color plate no. 21.—From the Hudson Bay watershed of North Dakota through the Great Lakes drainage, including southern Ontario and New York as far east as the upper tributaries of the St. Lawrence River; southward to western Pennsylvania, the Cumberland River system of Tennessee and to northern Kansas, Wyoming and Colorado. Intergrading southward with *I. m. catulus*, which occurs in the Gulf states as far southwest as Texas and (as native?) to northern Mexico. Locally introduced in the Pacific drainage, and elsewhere. Typically a fish of ponds, sloughs and the sluggish parts of creeks and rivers; in shallow and often silty water.

NORTHERN BROWN BULLHEAD—*Ictalurus nebulosus nebulosus* (LeSueur). Fig. 171 and color plate no. 22.—In southern Canada from as far west as southeastern Saskatchewan and from North Dakota and the Great Lakes—St. Lawrence region southward to the northern part of the Ohio Valley and from Nova Scotia to Virginia. Extensively introduced in western North America. Represented from southern Illinois, from northeastern Oklahoma and eastern Arkansas, to the Carolinas and Florida by *I. n. marmoratus* Holbrook (mottling in color plate 22 resembles *marmoratus* closer than *nebulosus;* identification of plate as given is tentative). Commonly in the somewhat weedy and deeper waters of lakes and sluggish rivers.

NORTHERN YELLOW BULLHEAD—*Ictalurus natalis natalis* (LeSueur). Fig. 172.—From the James River in North Dakota through the Great Lakes—St. Lawrence drainages southward to eastern Oklahoma and Texas and to the Tennessee River system (the form in the Hudson River system in New York is not placed subspecifically; the one occurring coastwise from New Jersey southward through the Austroriparian belt has been regarded as a different subspecies, *I. n. erebennus* (Jordan)). In sluggish to fairly rapid waters more common than the other bullheads in clean waters and in vegetation.

FLATHEAD CATFISH—*Pylodictis olivaris* (Rafinesque). Fig. 173.—From South Dakota, tributaries of Lake Michigan in Michigan, Lake Erie (where rare) and western Pennsylvania, south in the Mississippi Valley to the Gulf slope of Alabama, the Río Pánuco system of northeastern Mexico and the Arkansas River in Oklahoma. Usually a large-river form; young commonly under stones on riffles.

STONECAT--*Noturus flavus* Rafinesque. Fig. 174—From Manitoba and Wyoming and northeastern Colorado through the southern two-thirds of Wisconsin and of Michigan, to southern Ontario and to Lake Ontario, upper St. Lawrence River in Quebec, Lake Champlain, and the Hudson River (perhaps a canal immigrant) and the Allegheny River basins of New York; southward in the Gulf drainage to the Tennessee River system of Alabama and the Arkansas River drainage of Oklahoma (wrongly attributed to Texas). Generally on riffles of larger creeks and smaller rivers, but common also in Lake Erie and Saginaw Bay.

TADPOLE MADTOM—*Schilbeodes gyrinus* (Mitchill). Fig. 175.—From the Red and Assiniboine river systems of Manitoba and from all but the northernmost of the Red River system of North Dakota (and reportedly from Montana) through the entire Mississippi River and Lake Michigan drainages of Wisconsin and the Lake Michigan and Lake Huron tributaries in Michigan to Lake Ontario and upper St. Lawrence River, and their tributaries, of Ontario and New York, to the St. Lawrence River drainage in Quebec, and, possibly as introductions, to local areas in New Hampshire and Massachusetts; southward, on both sides of the Appalachian Mountains, to Florida and Texas, avoiding the Ozark Upland; introduced in the Snake River system of Idaho and Oregon. Characteristically in weedy habitats, with moderate to sluggish current and muddy bottom. This species was listed as S. *mollis* (Hermann) in the previous edition of this book.

NOTE—*Schilbeodes nocturnus* (Jordan and Gilbert) (Fig. 176) was included in the last edition of this book, on the basis of one specimen, from Lake Erie, that is now regarded as a hybrid, S. *gyrinus* × S. *miurus* (*fide* M. B. Trautman).

COMMON EASTERN MADTOM—*Schilbeodes insignis insignis* Richardson. Fig. 177.—From New Hampshire (possibly introduced) and from some southern tributaries of Lake Ontario in New York, southward, on the eastern slope, to northeastern Georgia (recent collections from the upper tributaries of the Tennessee River in Virginia are thought to represent an introduction). Usual habitat: under stones on riffles of creeks and rivers. This species was listed as S. *m. marginatus* (Baird) in the previous edition of this book. Represented in the Upper Kanawha River system and, less typically, in the Roanoke River system, with intergrades on either side of the Roanoke, by a subspecies, S. *i. atrorus* Hubbs and Raney.

SLENDER MADTOM—*Schilbeodes exilis* (Nelson). Fig. 178.—From central Iowa and southeastern parts of Minnesota and Wisconsin, and questionably, on the basis of an early record from southeastern Michigan, south to northeastern and southern Missouri, east-central Kansas, the Arkansas River drainage of eastern Oklahoma, and central Arkansas. Also in the Duck, Cumberland, and middle Tennessee river systems of Tennessee and the Tennessee River drainage of northern Alabama. One record (doubted by some) from the Guyandotte River in West Virginia. Rare and perhaps becoming extirpated toward the north and the east; abundant in the Ozark region (of Missouri, southeastern Kansas, eastern Oklahoma and northwestern Arkansas). Doubt has been attached to the records from the Great Lakes basin, in the southeastern parts of Wisconsin and Michigan. This species was given as S. *insignis* (Richardson) in the previous edition of this book.

BRINDLED MADTOM—*Schilbeodes miurus* (Jordan). Fig. 179.—From the eastern and southern drainages of Illinois to Lake Erie and its tributaries in Michigan and Ontario, to the Lake St. Clair drainage of Ontario, and to southern tributaries of Lake Ontario and to the Allegheny drainage in New York; south, largely avoiding the major uplands, to Mississippi, Louisiana, Arkansas, and southeastern Missouri, and to the Arkansas River system of eastern Oklahoma, southeastern Kansas, and southwestern Missouri. Generally in riffles of faster waters but sometimes in quiet weedy places and sand-clay habitats; rare on muddy bottoms.

FURIOUS MADTOM—*Schilbeodes furiosus* (Jordan and Meek). Fig. 180.—From the head of the Detroit River and the Lake Erie drainage of Michigan and western Ohio to the Ohio River tributaries of Indiana, Ohio, extreme western Pennsylvania, and central Kentucky. Under stones in riffles of medium-sized, clear to rather silty streams. This species was called S. *eleutherus* (Jordan) in the previous edition of this book and under that name covered a complex of two species, one (outside the Great Lakes) as yet unnamed pending publication of studies by W. R. Taylor.

Mudminnow Family—*Umbridae*

(Fig. 181 and col. pl. no. 23)

Mudminnows are readily distinguishable from all other fishes of the region by the dark vertical bar at the base of the rounded tail fin. The scales are cycloid and have a unique surface sculpture.

Only three species of mudminnows are named. In addition to *Umbra limi* there are *Umbra pygmaea* (De Kay) of the lowland fresh waters along the Atlantic Coast and *Umbra krameri* Walbaum of southeastern Europe. A Western genus (*Novumbra*) and the Arctic blackfishes (*Dallia*) are referred to distinct though related families (or subfamilies).

The mudminnow, as its name implies, lives in soft-bottomed, sluggish or even stagnant habitats. It is extremely tolerant of adverse environmental conditions, including oxygen deficiency. Reportedly it can be superficially frozen in ice and will revive upon thawing. The mudminnow is often the only species to survive a severe winter kill. A typical escape habit of the fish is to dive into the bottom mud, from whence the common name. The spawning season is in the early spring and the eggs are laid on vegetation. This fish is apparently omnivorous, although principally carnivorous, feeding upon insect larvae and minute crustacea in addition to a certain amount of plant material. The scales are apparently not adapted to age determination. Growth, however, has been shown to be slow. Two or more years are required to attain the usual length of a few inches.

The economic significance of the mudminnow lies chiefly in its value as forage for predacious fishes and in its use as bait. The dark color which is thought by some to render it unsuitable as bait may be partly overcome by keeping the fish over light background in bright light.

Central mudminnow—*Umbra limi* (Kirtland). Fig. 181 and col. pl. no. 23.—Manitoba through almost the entire Great Lakes region (including Lake Nipissing) to the upper St. Lawrence basin of southern Quebec and the Lake Champlain basin (one record from the Erie Canal in the Mohawk River watershed); south, in the central basin of the continent, to the upper Ohio system, Reelfoot Lake in northwestern Tennessee, northeastern Arkansas, Kansas and possibly eastern Nebraska. Typically in spring-fed pools with a soft bottom, and often common in very stagnant situations.

Pike Family—*Esocidae*

(Figs. 182–188 and col. pl. no. 24)

Members of this family are distinguished by the duck-bill shape of the front of the head when viewed from above. Besides this, they possess stout, sharp teeth, have scales on the head and are covered with cycloid scales that are deeply scalloped on their anterior margins.

The pikes represent another Holarctic family, with the one genus *Esox*. The northern pike, *Esox lucius,* occurs in the northern parts of Eurasia and North America. In eastern Siberia there is another species, the black spotted *Esox reicherti* Dybowski. The North American species are listed below.

In the Great Lakes region, all members of this family, excepting the grass pickerel, are important game fishes. Because of their wide distribution and popularity, many local common names have been applied to them. The keys subsequently provided will help to distinguish the various forms and the geographic range will further aid in correlating local names with the ones which we have adopted for this text. All members of the group are predacious fishes, dwelling in the open waters of lakes, ponds and streams. All are essentially spring spawners and broadcast the eggs over the vegetation in shallow waters, usually in swampy or marshy areas. They are principally piscivorous, but include in their fare an occasional small muskrat, mouse, duckling or frog.

The northern pike enters the commercial fishery. The sport fisheries are widespread and very important, being particularly good for "muskies" in larger lakes with deeper channels and for the northern pike in shallower waters, such as newly formed shallow impoundments. The overall best and least expensive lures for these fishes are flashy spoons with or without colors. Live minnows also make excellent decoys. Northern pike are readily taken by spearing through the ice after having been decoyed by live or artificial fish used as bait. For the northern pike, sizes range up to more than 54 inches and 35 pounds and for the muskellunge to more than 60 pounds. It is anomalous that the grass pickerel, which seldom exceeds a length of 14 inches, is ordinarily protected by the 14-inch legal size limit set for northern pike throughout most of this region, because most anglers do not distinguish the species. The grass pickerel, however, is principally a fish of the southern part of the Great Lakes area and is a common inhabitant of the smaller streams. All of these fishes show affinity for such cover as submerged stumps or logs, from which they are readily raised.

1 { Opercles wholly scaled (Fig. 39A) 2
Opercles scaleless on lower half (Fig. 39B, C) 3

2 { Adult with dark bands; branchiostegals (Fig. 40) usually 11 to 13; scales in lateral line about 105; size small (to 14 inches).
WESTERN GRASS PICKEREL—*Esox americanus vermiculatus* LeSueur. (Fig. 182 and col. pl. no. 24)
Adult with dark chain-like reticulations; branchiostegals usually 14 to 16; scales in lateral line about 125; size larger (to 2 feet).
CHAIN PICKEREL—*Esox niger* LeSueur. (Figs. 183, 184)

3 { Markings (except in young) in form of light spots; branchiostegals usually 14 to 16 (Fig. 40); sensory pores of head large, typically 5 on each side of mandible (Fig. 40); cheek fully scaled (Fig. 39B).
NORTHERN PIKE—*Esox lucius* Linnaeus. (Figs. 185, 186)
Markings in form of dark spots or bars; branchiostegals usually 17 to 19; sensory pores of head minute, usually 6 to 9 on each side of mandible; cheek usually scaleless on lower half (Fig. 39C).
GREAT LAKES MUSKELLUNGE—*Esox masquinongy masquinongy* Mitchill. (Figs. 187, 188)

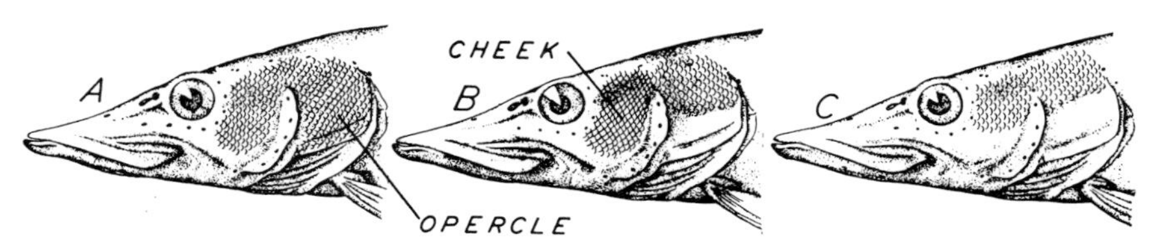

Fig. 39. Diagrams of heads of members of the pike family to show approximate extent of squamation on cheeks and opercles. A. Cheek and opercle wholly scaled, western grass pickerel (*Esox americanus vermiculatus*). B. Cheek fully scaled, upper half of opercle only scaled, northern pike (*E. lucius*). C. Cheek and opercle both scaled only on upper half, Great Lakes muskellunge (*E. m. masquinongy*).

WESTERN GRASS PICKEREL—*Esox americanus vermiculatus* LeSueur. Fig. 182 and col. pl. no. 24.—From southeastern and reputedly from southwestern Wisconsin, and from eastern Iowa through Illinois to the southern half of the Lower Peninsula of Michigan, southern Ontario, upper St. Lawrence River (rare) and Lake Champlain; southward, from the upper Mississippi Valley (Nebraska to Pennsylvania) to the Gulf Coast from

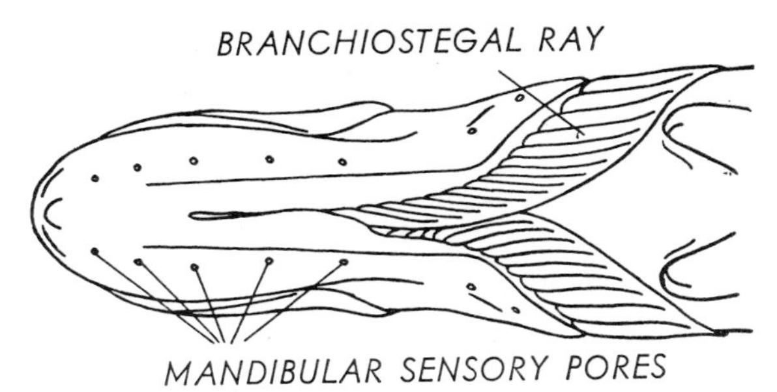

Fig. 40. Lower surface of head of northern pike (*Esox lucius*), showing 15 branchiostegal rays on left side (upper in figure) and 5 sensory pores on each side of mandible (lower jaw).

Alabama to eastern Texas (the southern form or forms are perhaps subspecifically distinct, and all forms are now thought, without final proof of geographic intergradation, to be subspecies of the Atlantic Coast form, *E. a. americanus* Gmelin, which ranges from eastern Florida to southern Canada); generally avoiding the foothill and mountain regions. Usually in quiet, weedy waters, over a mud bottom. (Not to be confused with the "grass pike," a name frequently used for the northern pike, *E. lucius.*)

Chain pickerel—*Esox niger* LeSueur. Figs. 183 and 184.—From Nova Scotia (introduced?), New Brunswick and the St. Lawrence River system of Quebec and the Lake Ontario drainages (at least in New York) southward, east of the mountains, to Florida; thence, in the Mississippi Valley (perhaps as a distinct subspecies), to northeastern Texas, southern Missouri and the Tennessee River system in Alabama; introduced into the Lake Erie drainage of New York. In quiet, weedy waters.

Northern pike—*Esox lucius* Linnaeus. Figs. 185 and 186.—Northern parts of the Northern Hemisphere; in North America from Arctic Alaska and the Pacific drainage in northern British Columbia to the Ungava Bay drainage and to Labrador, south to northern New England, the Hudson River drainage of eastern New York, the northern part of the Ohio Valley, the entire Great Lakes district, Missouri, eastern Nebraska, and the Peace River system of British Columbia. In cool to moderately warm, generally weedy, lakes, ponds and sluggish rivers. Whether the North American form (or forms) is different from those recognized in Eurasia has not been ascertained.

Great Lakes muskellunge—*Esox masquinongy masquinongy* Mitchill. Figs. 187 and 188.—Scatteringly distributed in the Great Lakes and St. Lawrence basins, including southwestern Quebec and Lake Champlain; and in some lakes of Ontario north to Abitibi and westward to Lake of the Woods. Most frequently caught in medium-sized to large lakes and in quiet parts of rivers; typically in clear, cool, weedy waters. Becoming scarce.

Other subspecies are *E. m. ohioensis* Kirtland, from the Ohio–Tennessee River system, including Chautauqua Lake, New York, an unnamed but possibly recognizable subspecies from some lakes of Ontario north to Lake Abitibi and westward to Lake of the Woods, and *E. m. immaculatus* Garrard, from headwaters of the Mississippi. Subspecies *masquinongy* seems to be typical only from the Lake St. Clair–Lake Erie area eastward; specimens from the Lake Huron and the Lake Michigan basins approach *E. m. immaculatus* and may represent still another form.

FRESHWATER EEL FAMILY—*Anguillidae*

(Fig. 189 and col. pl. no. 25)

True eels can be told from all other northern freshwater fishes, except lampreys, by the snakelike shape and movements. As already indicated, they may be easily distinguished from lampreys by the true jaws which are set with ordinary teeth and by the presence of pectoral fins and scales. The structurally unique cycloid scales of the freshwater eels, however, are embedded in the skin and can be seen only by close inspection.

In this family of eels there are about sixteen species. The European eel is *Anguilla anguilla*. Other species occur in the vast Indo-Pacific fauna. All spawn in the sea but live and grow through most of their lives in fresh and brackish waters. There are no representatives of this family in the eastern Pacific. There are many other eels, such as the conger eels, snake eels and morays, in tropical and subtropical seas around the world.

The American eel is found principally along the Atlantic and Gulf coasts of North America, penetrating the tributary waters. In the Great Lakes, it was native only in Lake Ontario. This eel enters fresh water to grow, but spawns only in the tropical Atlantic Ocean. In 1877 and subsequent years to 1891, it was introduced into Michigan waters. Until a few years ago, in landlocked lakes, specimens were still occasionally encountered that were, therefore, as much as fifty years old. Individuals stocked in landlocked inland waters are, of course, doomed eventually to die without spawning, since descent to the sea is necessary for reproduction. In other waters the stocked fish early disappeared, as they attempted to migrate to the sea. The few eels recently taken in the upper Great Lakes presumably had come in by way of canals. Permanent establishment in the upper Great Lakes is of course impossible.

Size attained may be as much as six feet, but the majority are less than half this length. Males are much smaller than females and both die following spawning. Eels are reported to be the most voracious of all carnivorous fishes. They are principally nocturnal feeders. The white flesh of the eel is esteemed, both when fresh and when smoked or otherwise cured.

AMERICAN EEL—*Anguilla bostoniensis* (LeSueur). Fig. 189 and color plate no. 25.—Eastern North America and (rarely) northern South America, from southern Greenland and Labrador to Brazil; in the Mississippi Valley as far inland as Colorado, Kansas, Nebraska, South Dakota, Minnesota, Wisconsin, Illinois, Indiana, Ohio and Pennsylvania; not native to the Great Lakes above Niagara Falls, but formerly introduced and abundantly established, and perhaps still extant in a few landlocked lakes; also (very rarely) in lakes Huron and Erie (reached no doubt *via* canals).

KILLIFISH FAMILY—*Cyprinodontidae*

(Figs. 190–193 and col. pl. no. 26)

Members of this family are all small soft-rayed fishes, with scales (cycloid in Great Lakes species) on the head as well as body. Most species, including all in the Great Lakes region, have the head flattened on top toward the snout and have a mouth that opens along the upper surface of the head. These are adaptations to the distinctive surface-feeding habits of the group. Furthermore these species are markedly barred or striped.

Killifishes occur in the temperate and tropical parts of North and South America, throughout most of Africa, in Spain and in southern Asia and the East Indies. There are scores of species that inhabit fresh, brackish, mineralized, and salt waters of the New World. In North America the group is best developed in and near Florida.

The killifishes are closely related to the livebearers, although they do not bring forth their young alive. They are common inhabitants of shoal waters of the Great Lakes and many of the larger lakes and streams of the region. Some are surface feeders and all have value as forage fish and as live bait for anglers.

1
- Dorsal fin originating distinctly in advance of anal fin; body with crossbands in both sexes (*Fundulus diaphanus*) 2
- Dorsal fin originating distinctly behind front of anal fin; body with crossbands in male only 3

2
- *Scales 40 to 55, usually 45 to 49; dorsal rays 12 to 15, usually 13 or 14; anal rays 10 to 13, typically 11 or 12; pectoral rays 15 to 19, usually 16 or 17; bars narrower, more intense and regular in shape; bars on caudal peduncle often short but not fused into a median lengthwise stripe; scale pockets sharply outlined with dark lines; extensions of anterior lateral bars across back more or less intact (obsolete in large specimens); anterior bars in males 9 to 15, the larger specimens with the higher number; size larger (to 110 mm. standard length, commonly more than 70 mm.).

  EASTERN BANDED KILLIFISH—*Fundulus diaphanus diaphanus* (LeSueur). (Fig. 190)
- *Scales 35 to 49, usually 40 to 44; dorsal rays 10 to 14, usually 11 to 13; anal rays 9 to 12, typically 10 or 11; pectoral rays 13 to 17, usually 14 or 15; bars broader, more diffuse and less regular in shape; bars on caudal peduncle generally fused into a median lengthwise stripe; scale pockets with diffuse dark borders; extensions of anterior lateral bars across back usually disrupted into irregular spots (obsolete in large specimens) (Fig. 191); anterior bars in males 5 to 10, the larger specimens with the higher number; size smaller (to 74 mm. standard length, very seldom more than 70 mm.).

  WESTERN BANDED KILLIFISH—*Fundulus diaphanus menona* Jordan and Copeland. (Fig. 191 and col. pl. no. 26)

3
- Body rather deep and compressed, depth 3.5 to 4.3 times in standard length; males with vertical bars, superimposed on horizontal streaks, females with about ten horizontal streaks; a black blotch below eye; median fins without dark speckles.

  NORTHERN STARHEAD TOPMINNOW—*Fundulus nottii dispar* (Agassiz). (Fig. 192)
- Body slender and scarcely compressed anteriorly, depth 4.4 to 5.3 times in standard length; a purplish black lateral band, with irregular edges in the male; no black blotch below eye; median fins with dark speckles.

  BLACKSTRIPE TOPMINNOW—*Fundulus notatus* (Rafinesque). (Fig. 193 and col. pl. no. 27)

BANDED KILLFISH—*Fundulus diaphanus* (LeSueur)—The range of this species is that of the two subspecies given below.

EASTERN BANDED KILLIFISH—*Fundulus diaphanus diaphanus* (LeSueur). Fig. 190.—From South Carolina north to the Maritime Provinces and Newfoundland; west through the eastern parts of Pennsylvania and New York including the Lake Champlain basin; locally in the Allegheny River system of Pennsylvania (doubtless through introduction). Lakes, quiet rivers and estuaries.

* This comparison was drawn up by Sidney Shapiro, on the basis of material from throughout the range of the species. In the Lake Ontario basin *F. d. diaphanus* intergrades with *F. d. menona*.

WESTERN BANDED KILLIFISH—*Funduluus diaphanus menona* Jordan and Copeland. Fig. 191 and col. pl. no. 26.—From the eastern parts of the Dakotas and northern Iowa through Minnesota, Wisconsin and Michigan and through the northern parts of Illinois, Indiana and Ohio to southern Ontario (Sudbury district and southward) and westernmost New York. Intergrading with *F. d. diaphanus* in the Lake Ontario and St. Lawrence basins. Apparently preferring shallow, quiet waters; most abundant in thin growths of rushes.

NORTHERN STARHEAD TOPMINNOW—*Fundulus nottii dispar* (Agassiz). Fig. 192.—From the Mississippi River tributaries of Iowa and southeastern Wisconsin and from the Lake Michigan drainage of Michigan and Indiana southward to northern Arkansas, eastern Oklahoma, and western Tennessee (replaced farther south, from the Carolinas to Texas, by other subspecies). Usually near the surface of clear, weedy backwaters.

BLACKSTRIPE TOPMINNOW—*Fundulus notatus* (Rafineque). Fig. 193 and color plate no. 27.—From Iowa and southeastern Wisconsin (both sides of the divide) to southern Michigan and the prairie regions of western and central Ohio; south to Kentucky, the Duck River of Tennessee and the Gulf drainages from Mississippi to Texas (distinct subspecies?), and to Oklahoma, Missouri and Kansas. Almost always at the surface of marginal waters in lakes and streams, where the current is lacking or moderate.

LIVEBEARER FAMILY—*Poeciliidae*

(Figs. 194 and 195)

Members of this family are small fishes that resemble the killifishes but they bear their young alive. This habit is best known for a tropical species, the guppy (*Lebistes*), which is kept by thousands of aquarists. Gravid females are instantly recognizable by the young which they contain and adult males are marked by the conspicuously elongated anal fin. The modified anal fin is used as an intromittent organ by the male to guide spermatozoa into the female. Fertilization of the eggs takes place inside the mother. All other fishes of the region are oviparous (producing eggs that are fertilized after leaving the female).

This family is exclusively American, ranging from the upper Mississippi Valley to Argentina. Species are particularly numerous in Mexico and Central America and in Cuba.

Gambusias or "mosquitofish" were introduced into Michigan during the early 1940's. This introduction was part of a northward acclimatization experiment of this preeminently southern fish. Certain of the introduced populations appears to have become well established. Their surface feeding habit makes them particularly valuable in the control of mosquitos. Mosquito larvae, "wrigglers," suspend themselves from the surface film of the water and are ready prey for gambusias. The value of this fish in control of malaria is well confirmed.

COMMON GAMBUSIA—*Gambusia affinis* (Baird and Girard).—In its native form this species occupies the ranges of the two widely distributed subspecies delimited below, plus the ranges of additional subspecies in Texas and northeastern Mexico, as far south as the Río Pánuco system.

WESTERN COMMON GAMBUSIA—*Gambusia affinis affinis* (Baird and Girard). Figs. 194 and 195.—Southern Illinois and southern Indiana to Alabama and the mouth of the Rio Grande; intergrading in Alabama with *G. a. holbrookii* of the Atlantic Coast; widely introduced around the world for mosquito control. Stocked and established in various localities about Chicago and in lower Michigan. Intergrades between *G. a. affinis and G. a. holbrookii* have also been introduced into southern Michigan but the stock did not become permanently established.

EASTERN COMMON GAMBUSIA—*Gambusia affinis holbrookii* Girard.—Eastern Alabama and Florida to New Jersey. This subspecies is known from the Great Lakes basin only from a temporary establishment of *affinis* × *holbrookii* intergrades in the Huron River system of southern Michigan.

Cod Family—*Gadidae*

(Fig. 196)

The burbot, a freshwater representative of this marine family, is easily told from all other fishes of the region by the single, prominent barbel on the underside of the chin near its tip. The small, cycloid, imbedded scales are unlike those of any other freshwater fish.

The family Gadidae comprises several score species, ranging from Arctic to Antarctic seas, with relatively few, mostly deepwater forms, within the Tropics. The freshwater species is Holarctic. The typical subspecies, *Lota lota lota* (Linnaeus), occurs in the northern parts of Eurasia.

The burbot is a fish of northern waters. In the Great Lakes region it lives chiefly in the Great Lakes proper, but occurs also in cooler streams, both sluggish and swift. It is taken to the great depths of about 700 feet, where it feeds upon the "chubs" of the whitefish family. It is a winter spawner. A length of thirty inches or more is attained although the average is about half this length. The usual weight is about one pound. The burbot enters the commercial catch only sparingly and in this region has only a very small market value, although elsewhere it is highly esteemed and brings good returns to fishermen.

American burbot—*Lota lota lacustris* (Walbaum). Fig. 196.—From the eastern portion of the Hudson Bay drainage and the Ungava Bay drainage of northeastern Canada south to the Connecticut, Delaware and Susquehanna systems, and all the Great Lakes basins; in the Missouri River system south to Missouri, Kansas and Wyoming; in the Mississippi River and tributaries throughout Minnesota; also in the Columbia River watershed; probably intergrading in the Fraser and Mackenzie river systems and the western portion of the Hudson Bay drainage with *L. l. leptura* Hubbs and Schultz (of the Yukon River region, other waters in Alaska and northeastern Siberia). In cool waters of large to small lakes and medium to large streams. In lakes chiefly in deep water; in streams preferring patches of plants and trash when young, stony riffles when half-grown and undercut banks when adult.

Troutperch Family—*Percopsidae*

(Fig. 197)

Troutperch are spotted fish with both spiny-rayed and adipose dorsal fins, weakly ctenoid scales of a unique kind and toothed jaws. The troutperch is particularly characteristic of the shoal waters of the Great Lakes proper and a few of the larger inland lakes of the area. It also occurs in a few streams. The fish is a particularly obvious part of the fauna when one collects at night. The troutperch spawns in the spring, using the tributary streams and probably the lake shallows for this purpose. It is a forage fish for predacious food fishes and seldom exceeds six inches in length.

There are only two species in this family, the one treated below and *Columbia transmontana* Eigenmann and Eigenmann of the Columbia River system. These are obviously relicts of a group that is now largely extinct. They are noteworthy for the combination they display of the characters of the soft-rayed and the spiny-rayed fishes.

Troutperch—*Percopsis omiscomaycus* (Walbaum). Fig. 197.—Canada from the Yukon River system and from the Mackenzie River basin (from the Arctic Circle to British Columbia), eastward to Hudson Bay and Quebec; throughout the Great Lakes–St. Lawrence system to Lake Champlain; south on the Atlantic slope (generally rare) from the Hudson River system to the Potomac River, and in the Mississippi Valley to West Virginia, Kentucky, Missouri, Kansas and South Dakota (rare and local southward). In all the Great Lakes proper, including the marginal lakes that were flooded by the Glacial Great Lakes; occurring in the interior also, from the western part of the Upper Peninsula of Michigan and Isle Royale northward, but seldom living in streams or inland lakes in the southern part of the Great Lakes basin (exceptional records are for the Manistee River in Michigan, and for the Genesee River above the falls and the Barge Canal in New York); north, east, south and west of the Great Lakes occurring commonly in silty rivers and creeks.

PIRATEPERCH FAMILY—*Aphredoderidae*
(Fig. 198)

The pirateperch is marked by the dislocation of the anus from its ordinary position at the front of the anal fin to the throat region. It is a spiny-rayed fish with ctenoid scales of a unique type.

The pirateperch is far from abundant in the Great Lakes region. It is particularly an inhabitant of creeks and is sometimes found in larger rivers and lakes. Its food is composed of small fishes and aquatic insects and their larvae. Spawning is in the spring and the parents are reported to guard the nest and young after hatching. The size seldom exceeds four inches.

There is only one living species in this family. All near relatives are extinct and are known only from fossil remains. The troutperches are distant cousins.

WESTERN PIRATEPERCH—*Aphredoderus sayanus gibbosus* LeSueur. Fig. 198.—From southeastern Minnesota, southern Wisconsin, southern Michigan and the southern tributaries of Lake Ontario southward, west of the Appalachian Mountains and foothills, to the Gulf Coast, as far as southeastern Oklahoma and Texas. A lowland form: northward in densely weedy creeks; southward chiefly in muddy sloughs and swamps. Replaced in the Atlantic drainage by *A. s. sayanus* (Gilliams).

BASS FAMILY—*Serranidae*
(Figs. 199 and 200)

Serranids are spiny-rayed fishes with ctenoid scales. In the region of the Great Lakes, two of the three species that represent the family are distinguished, except as young, by prominent, horizontal dark lines on the upper part of the sides.

Most Serranidae are tropical and subtropical marine fishes. The group is a large and diverse one, composed of groupers, jewfishes and many other types. There are scores of marine species but only a few that breed in fresh waters. Some of these, including the striped bass, *Roccus saxatilis* (Walbaum), of food and game repute, live in the sea except during spawning time.

In the Great Lakes the white bass is the only member of this family that is sufficiently common to merit mention here in regard to natural history. Its general habitat is deep, quiet water over sand and gravel bottoms, in medium or large lakes and in larger deep rivers. The fish is notably gregarious, often schooling at the surface, as many anglers have learned.

Spawning is in the spring in shallow, sometimes moving water. The food of juveniles and adults is principally fish and crustaceans. Growth is reported to be extremely rapid. Young-of-the-year attain more than six inches in length by the first fall of life. The white bass is a good game and food fish, having both a commercial fishery value and a recreational value. The best method of fishing for white bass appears to be by the use of live minnows for bait, with the aid of a lantern or other light when fishing after dark. It also takes artificial flies or spinners. Sizes are reported to range up to eighteen inches in length and three pounds in weight.

1. Dorsal fins separated; formula for anal fin III, 11 to 13; the anal spines graduated, the first about half the length of the second, and second distinctly shorter than third; lower jaw projecting; base of tongue with teeth.

   WHITE BASS—*Roccus chrysops* (Rafinesque). (Fig. 199)

   Dorsal fins slightly conjoined at base; formula for anal fin III, 10; the anal spines not graduated, the first scarcely one-third as long as second, the second and third subequal; jaws almost equal; base of tongue toothless 2

2. Spines stout; longest dorsal spine more than 0.5 in head; body striped with about 7 horizontal black stripes; sides commonly brassy in background color; dorsal fin with two slightly connected parts.
Yellow bass—*Morone interrupta* Gill. (Fig. 200)

Spines slender; longest dorsal spine about 0.5 in head; body evenly colored; background silvery; dorsal fin with two well-connected parts.
White perch—*Morone americana* (Gmelin)

White bass—*Roccus chrysops* (Rafinesque). Fig. 199.—From the Mississippi River system of southern Minnesota and of Wisconsin and probably from Lake Superior about the Apostle Islands, Wisconsin (an old report), and from the Michigan-Huron lake basin to southern Ontario (including Lake Nipissing); through New York in the Lake Ontario and St. Lawrence River basins and perhaps in the Mohawk River; down the St. Lawrence to the city of Quebec, southward, west of the mountains, to the Tennessee River drainage of Alabama and to Mississippi and part of eastern Texas and eastern Oklahoma. Rather widely distributed through the Great Lakes, chiefly near the shore of the main lakes, but apparently now lacking in Lake Superior and generally rare northward (where a marked depletion was noted about 1880). Ordinarily in the larger and more open rivers, impoundments, and lakes; often in large schools at the surface.

Yellow bass—*Morone interrupta* Gill. Fig. 200.—From the southern parts of Minnesota, Wisconsin and Indiana south to the Tennessee River drainage in Alabama and thence to Louisiana, eastern Texas and eastern Oklahoma. Reported for the Great Lakes drainage only once, from the head of the Chicago Drainage Canal. Generally in larger rivers.

White perch—*Morone americana* (Gmelin). Atlantic Coast of North America from New Brunswick, Prince Edward Island, and Nova Scotia to South Carolina. Introduced in various waters of New England, New York, Quebec (presumably, St. Lawrence River), the Lake Ontario basin, and Lake Erie. Canals may account for the recent entry into the drainages of lakes Erie and Ontario. In brackish and fresh waters, ascending streams, and landlocked in many ponds.

Perch Family—*Percidae*

(Figs. 201–227 and col. pls. 28–34)

The perch family is a diverse, spiny-rayed group which is bound together and distinguished from other fishes of the region by having two distinct dorsal fins along with only one or two spines in the anal fin. The scales are moderately to strongly ctenoid.

Three subfamilies make up this family, which in North America is native only to the waters east of the Continental Divide; fishes typifying each are: (1) yellow perch (Percinae); (2) walleyes (also called pikeperches) and saugers (Luciopercinae); (3) darters (Etheostomatinae). Members of the first two groups are inhabitants of the Great Lakes and inland lakes as well as larger streams. Most darters are stream dwellers but some, such as the Iowa darter, are pre-eminently lake or pond fishes. The Percinae and Luciopercinae, each with fewer than a dozen species, occur in Europe as well as in North America. The darters, exclusively North American, comprise about one hundred species.

All members of this family spawn in the spring so far as known; no species in the Great Lakes region breeds after early summer. The yellow perch lays a zigzag ribbon ("rope") of eggs in moderately shallow water. The fishes of the pikeperch group scatter their eggs on the bottom in shoal areas. The darters are diverse in their habits: (1) some, as the least and Iowa darters and the logperch, leave the eggs unguarded on the substratum; (2) others, such as the rainbow and orangethroat kinds, bury their eggs in fine gravel on riffles and desert them; but (3) still others, for example the Johnny darter and the fantail, practice parental care—the males establish nests beneath flat stones or other objects on the bottom and courageously guard the eggs, which are laid on the ceiling of the breeding cavity.

All fishes in this family are carnivorous and as young feed mostly on microscopic animals. With increasing size aquatic insects and their larvae and a host of kinds of other invertebrates are eaten. As they become larger, perch and members of the pikeperch subfamily eat other fishes, including minnows.

The walleye, often called pike, pickerel, or pikeperch, is the largest member of the perch family. Its maximum length is usually given as about three feet and weight, as about twenty-five pounds. The saugers and blue walleye are smaller, attaining only about half the length of the yellow walleye. Yellow perch are usually taken at lengths of less than fourteen inches and weights of less than a pound. Darters are the smallest; all are more or less of "minnow" size. Perhaps the smallest kind of fish in the region is the least darter, which matures sexually at sizes approximating one inch. Some darters are among the most brilliantly colored of the freshwater fishes.

The yellow perch is an important food and sport fish throughout the southern part of the Great Lakes region. It composes a significant part of the commercial catch and is taken in great numbers by anglers. The perch is caught by still-fishing with live bait–small minnows or worms. It is often taken in the winter through the ice and responds well to small minnows. Best catches are made along the lower reaches of streams entering the Great Lakes and in shallow parts of these lakes in the spring during the spawning runs.

In the Great Lakes yellow walleyes attain their greatest abundance in Lake Erie, where they make up a large part of the commercial catch. As game fish, they are more important throughout the northern part of the region in inland lakes and larger rivers. They are taken by deep trolling with live bait or with bright spoons. Bait casting with submersible lures also produces some, particularly at night, when they come into the shallows to feed, or at the bases of dams. They are also taken through the ice.

The saugers and blue walleyes make up a considerable and increasing proportion of the commercial take of fishes from Lake Erie, but they are not very important as game fishes in the Great Lakes region.

1 Preopercle strongly serrate (Fig. 42); mouth large, the upper jaw extending at least to below middle of eye; maxillary with posterior part of upper border not concealed by suborbital; branchiostegals 7 (rarely 8); pseudobranchia well developed; no distinct genital papilla; fishes of medium to large size, maturing (some stunted stocks of perch excepted) at a length of more than 6 inches 2

Preopercle with totally smooth edge (weakly serrate in some extralimital forms); mouth small, the upper jaw usually not extending to below middle of eye, and never much farther; maxillary with upper posterior edge slipping under suborbital; branchiostegals 6 (7 in some extralimital forms); pseudobranchia rudimentary or lacking; a distinct genital papilla (large in adult females); small fishes, not exceeding 6 or 7 inches, usually 1 to 4 inches long (darters, Etheostomatinae) 5

2 No canine teeth (Fig. 41B); body moderately deep and compressed; pelvic fins inserted close together; anal fin formula II, 6 to 8; body with definite crossbands.

YELLOW PERCH—*Perca flavescens* (Mitchill). (Figs. 3, 201 and col. pl. no. 28)

Canine teeth strong (Fig. 41A); body slender, nearly terete; pelvic fins rather widely separated (distance between bases approximately equals width of base of either fin); anal fin formula II, 12 or 13; crossbands rather indefinite or not apparent (*Stizostedion*) 3

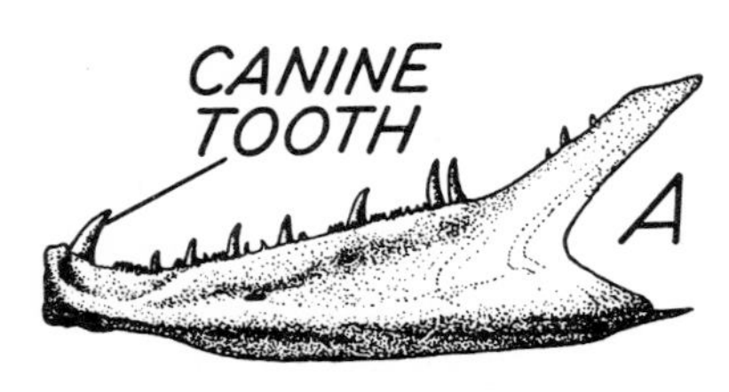

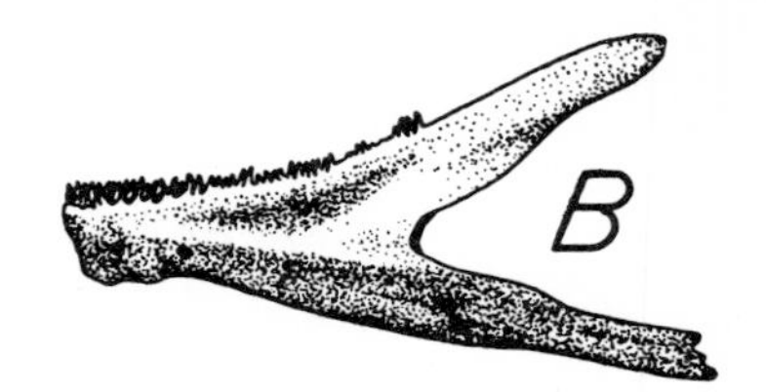

Fig. 41. Lower jaw bones of two members of the perch family, showing dentition. A. Canine teeth present (yellow walleye, *Stizostedion v. vitreum*). B. Cannine teeth absent (yellow perch, *Perca flavescens*).

3

Dorsal fins with rows of round, black spots (not evident in young); end of spinous dorsal without a prominent black blotch; young with color pattern in part longitudinal; cheeks as a rule closely scaled; rays of soft dorsal 17 to 20; pyloric caeca 5 to 8, the 4 longest much shorter than the stomach.

SAUGER—*Stizostedion canadense* (Smith). (Fig. 202)

Dorsal fins with obscure dusky mottlings; end of spinous dorsal with a large black blotch; young with color pattern transverse; cheeks as a rule sparsely scaled; rays of soft dorsal 19 to 22; pyloric caeca 3, each about as long as the stomach (*Stizostedion vitreum*) 4

4

Body in life with brassy yellow mottlings, never bluish; lower fins clear yellowish; eyes smaller and set farther apart (the bony interorbital width measures 1.1 to 1.4 in the length of the orbit in the young and half-grown, and is about equal to the orbit in adults).

YELLOW WALLEYE—*Stizostedion vitreum vitreum* (Mitchill). (Fig. 203 and col. pl. no. 29)

Body in life grayish blue, without brassy yellow mottlings; lower fins bluish white; eyes larger and set more closely (the bony interorbital width typically measures 1.4 to 2.0 times in the length of the orbit).

BLUE WALLEYE—*Stizostedion vitreum glaucum* Hubbs. (Fig. 204)

5

Flesh pellucid; scales of trunk confined to middle of sides; body very elongate (pencil-shaped; depth 7.8 to 9.0 times in standard length).

NORTHERN SAND DARTER—*Ammocrypta pellucida* (Baird). (Fig. 211)

Flesh opaque; squamation of body almost or quite complete; body less extremely elongate (depth 5.0 to 7.0 times in standard length) 6

6

Midline of belly with a definite single file of scales which are more or less enlarged, thickened, deciduous and separated by a slight groove from the scales on either side (Fig. 43) (these scales only weakly specialized in *Hadropterus shumardi* and sometimes little modified in females of other genera); one such scale between pelvic fins (Fig. 44); pelvics separated by a space at least three-fourths as wide as base of either fin; anal fin usually almost as large as second dorsal, sometimes even larger; body typically elongate and little compressed; vertebrae 37 to 44 7

Midline of belly without a median file of specialized scales; no specialized extra-spiny scale between pelvic fins; pelvic separated by a space less than three-fourths as wide as base of either fin (except in *Etheostoma chlorosomun* and *E. nigrum*); anal fin smaller in area than second dorsal fin; body typically deeper and more compressed; vertebrae 32 to 43 (*Etheostoma*) 13

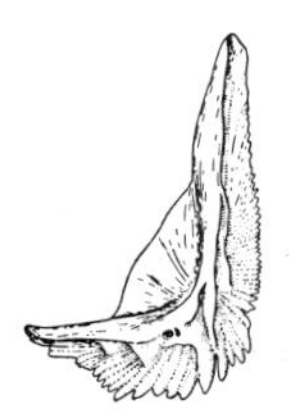

Fig. 42. Left preopercle of yellow perch (*Perca flavescens*), showing serrate (toothed) ventral and posterior margins.

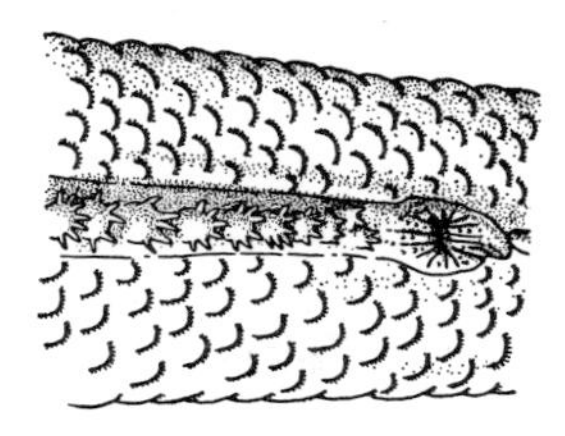

Fig. 43. Midline of belly with a definite single file of enlarged, caducous scales which are separated by a slight groove from the scales on either side (northern logperch, *Percina caprodes semifasciata,* male).

7 Premaxillaries distinctly protractile, even on the midline separated by a deep groove from the snout; air-bladder rudimentary (subgenus *Cottogaster*).
CHANNEL DARTER—*Hadropterus copelandi* (Jordan). (Fig. 209)

Premaxillaries scarcely protractile, bound down to the snout by a median fleshy frenum (which, however, may be crossed by a shallow groove, especially in *Hadropterus shumardi,* when the mouth is tightly closed); air-bladder rather large (but thin-walled), well developed in young 8

8 Snout not extended so far forward as the upper lip; mouth more or less oblique; anal spines stiff; scales fewer than 80; crossbars either broad or obsolete (*Hadropterus*) 9

Snout extended forward as a small conical protuberance beyond the upper lip; mouth horizontal; anal spines very flexible; scales 78 to 103 in lateral line; crossbars numerous and narrow (*Percina caprodes*) 12

9 Belly largely scaleless medially, but crossed before anus by a bridge of scales; scales of midline only incipiently modified; premaxillary frenum usually hidden by a furrow behind upper lip (subgenus *Imostoma*).
RIVER DARTER—*Hadropterus shumardi* Girard. (Fig. 205)

Belly mostly scaled, and with the scales of the midline strongly modified (at least in males); premaxillary frenum usually not hidden by a crossfurrow (subgenus *Hadropterus*) 10

10 Scales in lateral line 52 to 67; cheeks naked; bands broad, large and squarish; color gilt in life.
GILT DARTER—*Hadropterus evides* (Jordan and Copeland). (Fig. 208)

Scales in lateral line 60 to 85; cheeks scaled; bands narrower or smaller; color not gilt in life 11

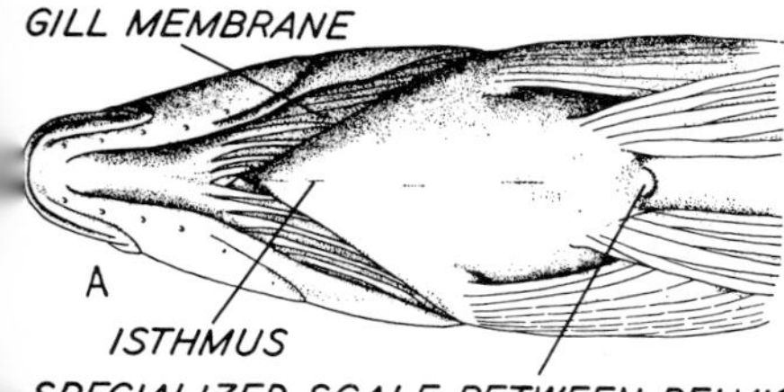

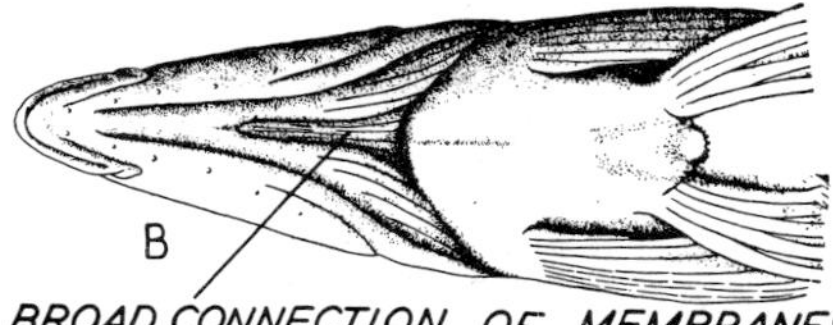

Fig. 44. Ventral view of anterior portions of two darters. A. Gill membranes of one side not connected with those of the other, snout rather blunt (blackside darter, *Hadropterus maculatus*). B. Gill membranes of one side rather broadly connected with those of the other (partly covering isthmus), snout sharp (slenderhead darter, *H. phoxocephalus*).

11

Lateral blotches large; spinous dorsal without a submarginal orange band; gill-membranes not connected (Fig. 44A); snout rather blunt.

Blackside darter—*Hadropterus maculatus* (Girard). (Figs. 44A, 206 and col. pl. no. 30)

Lateral blotches small; spinous dorsal with a submarginal orange band in life; gill-membranes rather broadly connected (Fig. 44B); snout sharp.

Slenderhead darter—*Hadropterus phoxocephalus* (Nelson). (Figs. 44B and 207)

12

Nape closely scaled; bars more even and regular, scarcely expanded into blotches in a median lateral row.

Ohio logperch—*Percina caprodes caprodes* (Rafinesque).

Nape with a trianguler scaleless area; bars less even, and often expanded into blotches in a median lateral row.

Northern logperch—*Percina caprodes semifasciata* (De Kay). (Fig. 210 and col. pl. no. 31)

13

Maxillaries fused to preorbital at sides (Fig. 45); premaxillaries overhung slightly by the very gibbous snout; color largely green in life.

Northern greenside darter—*Etheostoma blennioides blennioides* Rafinesque. (Figs. 226, 227 and col. pl. no. 34)

Maxillaries free from preorbital at sides; premaxillaries projecting forward as far as, or usually farther than, the more or less pointed snout; color not largely green in life (except often in *Etheostoma zonale*) 14

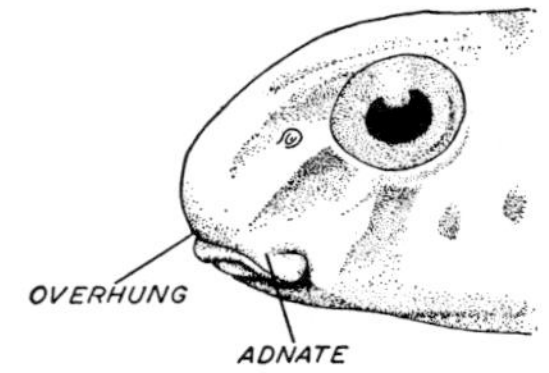

Fig. 45.

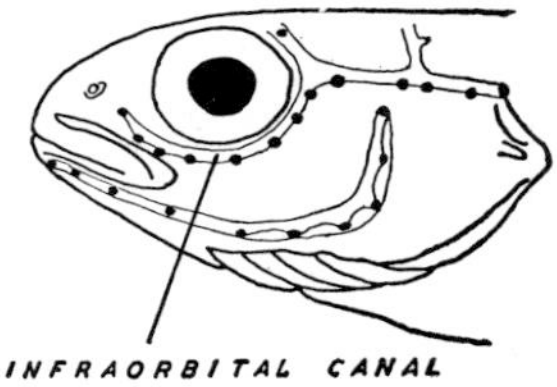

Fig. 46.

Fig. 45. Maxillaries fused to preorbital at sides; premaxillaries overhung slightly by the very gibbous snout (northern greenside darter, *Etheostoma b. blennioides*). (The eye is incorrectly located in this diagram.)

Fig. 46. Lateral line system on the head of a darter, showing canal and pores. In the species shown, the infraorbital canal is complete and has 8 pores. (After Hubbs and Cannon, 1935.)

14

Premaxillaries protractile, separated from the snout by a deep cross-furrow; one anal spine (in Great Lakes forms), flexible; pelvic fins separated by a space about as wide as base of either fin (subgenus *Boleosoma*) 15

Premaxillaries nonprotractile, bound to the snout by a fleshy median frenum; two stiff anal spines; pelvic fins inserted very close together 18

15

Lateral line lacking on posterior part of body; dorsal fins widely separated; snout very blunt and much rounded; dark streak in front of eye continuous around snout. (Occurrence in Great Lakes doubtful.)

Bluntnose darter—*Etheostoma chlorosomum* (Hay). (Fig. 214)

Lateral line nearly or quite complete; dorsal fins little separated; snout more or less sharp; dark streak in front of eye broken at front of snout (*Etheostoma nigrum*) 16

16
- Dorsal soft-rays usually 13 to 15; soft dorsal high and finely tessellated in adult males; body more robust and snout sharper; cheeks scaly.
  TESSELLATED JOHNNY DARTER—*Etheostoma nigrum olmstedi* Storer.
- Dorsal soft-rays usually 11 to 13; soft dorsal lower, and coarsely banded with dark; body usually slender and snout blunter 17

17
- Nape, cheeks and breast well scaled.
  SCALY JOHNNY DARTER—*Etheostoma nigrum eulepis* (Hubbs and Greene).
- Nape, cheeks and breast scaleless.
  CENTRAL JOHNNY DARTER—*Etheostoma nigrum nigrum* Rafinesque. (Figs. 212, 213)

18
- Lateral line obsolete; dorsal spines usually 6; lower fins much produced in breeding males; minute fishes, seldom as long as 1½ inches (subgenus *Microperca*).
  LEAST DARTER—*Etheostoma microperca* Jordan and Gilbert. (Figs. 224, 225)
- Lateral line developed, at least on trunk; dorsal spines usually 8 or more; lower fins not greatly produced; size commonly more than 1½ inches (the former genus *Poecilichthys*) 19

19
- Dorsal spines not ending in fleshy knobs; lower jaw scarcely or not at all projecting; head partially scaled and males brilliantly colored 20
- Dorsal spines ending in fleshy knobs (conspicuous only in breeding males); lower jaw projecting, or the jaws about equal; head completely scaleless and males without bright colors (subgenus *Catonotus;* fantail, *Etheostoma flabellare*) 23

20
- Gill-membranes very broadly connected (Fig. 44B); pectoral fin longer than head; color greenish in life; lateral line complete.
  EASTERN BANDED DARTER—*Etheostoma zonale zonale* (Cope). (Fig. 216)
- Gill-membranes at most very slightly connected (Fig. 44A); pectoral fin shorter than head; color not greenish in life; lateral line incomplete (not extending beyond end of dorsal fin base) 21

21
- Cheeks scaled.
  IOWA DARTER—*Etheostoma exile* (Girard). (Fig. 218 and col. pl. no. 32)
- Cheeks naked (or with a few embedded scales around the eye in Great Lakes forms) 22

22
- Infraorbital branch of lateral line canal complete (Fig. 46) (partial desiccation and magnification needed); gill-membranes slightly connected; bars nearly complete across posterior part of body; light stripe on spinous dorsal best developed posteriorly; males (in life) with red on anal and caudal fins.
  RAINBOW DARTER—*Etheostoma caeruleum* Storer. (Figs. 219, 220)
- Infraorbital branch of lateral line interrupted below eye, leaving an isolated group of 4 pores on preorbital region; gill-membranes not connected, slightly overlapping at angle; bars short, submedian; light stripe on spinous dorsal best developed anteriorly; males (in life) without red on anal and caudal fins.
  NORTHERN ORANGETHROAT DARTER—*Etheostoma spectabile spectabile* (Agassiz). (Fig. 221)

23
- Lengthwise rows of spots or dashes inconspicuous (fairly well developed in breeding males).
  BARRED FANTAIL—*Etheostoma flabellare flabellare* Rafinesque. (Figs. 222, 223)
- Lengthwise rows of spots or dashes conspicuous (even in females).
  STRIPED FANTAIL—*Etheostoma flabellare lineolatum* (Agassiz). (Col. pl. no. 33)

YELLOW PERCH—*Perca flavescens* (Mitchill). Figs. 4, 6, 201 and color plate no. 28.—Lesser Slave Lake of the Mackenzie basin and the Hudson Bay drainage of west-central and eastern Canada south to South Dakota and to the northern parts of Missouri, Illinois, Indiana, Ohio (introductions southward largely unsuccessful) and to western Pennsylvania; in coastwise streams from New Brunswick to South Carolina (records for Kansas, Texas, Oklahoma, and southern Alabama are presumed to be erroneous or due to introduction); widely introduced in the West. Common throughout the Great Lakes to depths at least as great as 25 fathoms. In varied habitats, usually of lakes, ponds, and quiet parts of streams.

SAUGER—*Stizostedion canadense* (Smith). Fig. 202.—From most of the Hudson Bay drainage, including the Saskatchewan, Red, and Assiniboine rivers, to New Brunswick (generally rare in the St. Lawrence River system, but common in Lake Champlain); southward, west of the Appalachians, to West Virginia, to the Tennesse River in Alabama, to eastern Oklahoma, to the Red River in Texas, to eastern Kansas, Nebraska, Wyoming, southwestern Iowa and Montana. Montana specimens, with a very small eye, have been distinguished as subspecies *S. c. boreum* (Girard). In the Great Lakes region abundant in Lake Erie and Saginaw Bay, less common in lakes Michigan and Huron and their main tributaries; only locally common in the Superior basin. Most often in somewhat silty rivers and large lakes.

WALLEYE—*Stizostedion vitreum* (Mitchill).—The total distribution of the walleye in North America is that of the two subspecies that follow.

YELLOW WALLEYE—*Stizostedion vitreum vitreum* (Mitchill). Fig. 203 and color plate no. 29.—From the Mackenzie River, Great Slave Lake, the Peace River system of British Columbia, the Saskatchewan River system and the Hudson Bay region to Labrador; southward on the Atlantic slope to North Carolina (apparently native in the coastal streams of North Carolina, but generally supposed to have been introduced farther north in the slope), and west of the mountains, to the Alabama River system of Georgia, to the Tennessee River drainage of Alabama, and to northern Arkansas and Nebraska. Common through the Great Lakes and many of the inland lakes and rivers of the basin; in Lake Erie chiefly to the westward.

BLUE WALLEYE—*Stizostedion vitreum glaucum* Hubbs. Fig. 204.—Lake Erie, particularly east of the islands. Frequenting deeper and cooler waters. The "blue pikes" of Lake Ontario, of lakes in the St. Lawrence River and Lake Huron drainages in Ontario and of Lake Winnipeg in Manitoba have been identified with this subspecies but they are probably not quite identical. They need further study.

RIVER DARTER—*Hadropterus shumardi* Girard. Fig. 205.—Lake Winnipeg basin of Manitoba, western and northern Ontario, the Mississippi drainage of Wisconsin, southern Lake Michigan, the Au Sable River and southward along the entire eastern margin of Michigan, and in the western Lake Erie region of Ontario; southward to the Tennessee River system in Alabama, the Red River in Arkansas and the streams of the coastal plain of Texas. Generally in channels of medium- to large-sized rivers, on soft to gravelly bottoms.

BLACKSIDE DARTER—*Hadropterus maculatus* (Girard). Fig. 206, col. pl. 30. —Southern parts of Saskatchewan and Manitoba and southwestern Ontario, and from North Dakota to the western part of the Lake Ontario drainage in New York; southward, west of the mountains, to the Alabama River system of Alabama, the Ozark region, and northeastern Texas. More or less evenly distributed through the basins of the Great Lakes, Superior excepted. Usually in weak currents of streams; in midwater when young and often when adult.

SLENDERHEAD DARTER—*Hadropterus phoxocephalus* (Nelson). Figs. 44B and 207.—The Mississippi River drainage from Iowa, southeastern Minnesota and northwestern Wisconsin to western Pennsylvania; southward to the Cumberland and Duck river systems of Tennessee, the Red River system in southern Arkansas and southeastern Oklahoma, and the Arkansas River tributaries in Oklahoma and Kansas (where particularly abundant). Known in the Great Lakes basin only from the Fox River system in Wisconsin. A riffle darter, seeming to prefer fine gravel.

GILT DARTER—*Hadropterus evides* (Jordan and Copeland). Fig. 208.—Northwestern Wisconsin (an isolated population); the Maumee River (tributary to Lake Erie); the Ohio River drainage from Illinois to New York; thence south to the Tennessee River system of North Carolina, northern Georgia and Tennessee; west of the Mississippi from Iowa to the Ozark region in Missouri and in the Arkansas River drainage of Arkansas. Typically on the open bottom of fast, stony riffles.

CHANNEL DARTER—*Hadropterus copelandi* (Jordan). Fig. 209.—From the entire eastern margin of the Lower Peninsula of Michigan (including Lake Huron to more than 5 fathoms) and from Lake Erie to the upper St. Lawrence and tributaries including Lake Champlain; southward, west of the mountains, to the Alabama River system in Alabama, to the Red River drainage in Arkansas and southeastern Oklahoma, and to northeastern Oklahoma and southeastern Kansas. In the Great Lakes region living in the main lakes and in the deep current of the lower parts of the principal tributaries; to the southward, often in medium-sized creeks; seldom on fast riffles.

LOGPERCH—*Percina caprodes* (Rafinesque).—The range of the species is indicated in the two distributional statements that follow.

OHIO LOGPERCH—*Percina caprodes caprodes* (Rafinesque).—From some waters in the southeastern parts of Wisconsin and of Michigan (as intergrades with *P. c. semifasciata*), and from the upper parts of Lake Erie tributaries in Ohio, through the Ohio Valley (from Illinois to Pennsylvania and New York and south to the Tennessee River system in Alabama). Most often on sandy to bouldery riffles of medium-sized streams.

NORTHERN LOGPERCH—*Percina caprodes semifasciata* (De Kay). Figs. 43, 210 and col. pl. no. 31.—From the Mississippi River system in northern Iowa and in Minnesota northward to the Churchill River system in Saskatchewan; eastward through the Hudson Bay drainage and the Great Lakes region to the St. Lawrence River basin in Quebec and Vermont, and to the Hudson River system in New York (with one old record of the species from the Susquehanna River system in Pennsylvania and a few records from the Potomac River system). In Lake Huron to 16 fathoms. Intergrading in southeastern Iowa and northeastern Missouri with *P. c. carbonaria* (Baird and Girard), which ranges from western Florida to Texas. Common in shallow water on gravelly bottoms in lakes and medium-sized streams; breeding on the gravel.

NORTHERN SAND DARTER—*Ammocrypta pellucida* (Baird). Fig. 211.—From southeastern Michigan and southwestern Ontario to Quebec near Montreal and to the Lake Champlain drainage in Vermont, southward to West Virginia and Kentucky. Occurring only in scattered localities in the Great Lakes–St. Lawrence basin. Living on and in the sandy bottom of the flowing sections of larger creeks and rivers and on the sandy shoals of a few lakes.

JOHNNY DARTER—*Etheostoma nigrum* Rafinesque.—The range of the species extends southward from the assigned ranges of the three subspecies here treated to Georgia and even Florida on the Atlantic slope and to eastern tributaries of the Gulf of Mexico. In the Great Lakes to depths of 23 fathoms.

TESSELLATED JOHNNY DARTER—*Etheostoma nigrum olmstedi* Storer.—From the Maritime Provinces to the Ottawa River and the eastern part of the

Lake Ontario drainage, especially at lower elevations; perhaps not intergrading with *E. n. nigrum;* southward, near the coast, to North Carolina. In quiet waters, and on the riffles of streams.

SCALY JOHNNY DARTER—*Etheostoma nigrum eulepis* (Hubbs and Greene). —Glacial lake districts of the Mississippi drainage in portions of Minnesota, northern Iowa, Wisconsin and northern Illinois; eastward through the Great Lakes basin, but only in and about the base-level lakes marginal to Lake Michigan in Wisconsin, Indiana and both peninsulas of Michigan, in Lake St. Clair and the Detroit River, and in the bays around the islands of Lake Erie in Ontario, Ohio and Pennsylvania. An isolated population occurs in the Neosho River drainage of southwestern Missouri. Commonly found in somewhat weedier situations than those preferred by *E. n. nigrum.*

CENTRAL JOHNNY DARTER—*Etheostoma nigrum nigrum* Rafinesque. Figs. 212 and 213.—Southern Canada from Saskatchewan and Hudson Bay to western Quebec, and the northern United States from Colorado and North Dakota to the southern tributaries of Lake Ontario (replaced at lower elevations in the eastern part of the Lake Ontario basin by *E. n. olmstedi*); southward, west of the Appalachian Mountains, to the Gulf drainage of Mississippi and Alabama, and to the Red River system of southeastern Oklahoma and the Arkansas River system of Arkansas, Oklahoma, and Kansas (in the headwaters of streams from the Maritime Provinces southward to North Carolina occur forms related to *nigrum* and in part perhaps to be identified as that subspecies or as intergrades between *nigrum* and *olmstedi*). Abundant throughout the Great Lakes region except in the habitats pre-empted by *E. n. eulepis,* and except in the eastern part of the Lake Ontario basin where *E. n. olmstedi* holds forth. Generally in quiet water, on more or less sandy bottoms; in lakes and various types and sizes of streams.

BLUNTNOSE DARTER—*Etheostoma chlorosomum* (Hay). Fig. 214.—Mississippi Valley from Iowa, extreme southeastern Minnesota, Illinois (with one record, in need of verification, from the Lake Michigan basin of Illinois), and southern Indiana to the Alabama River system of Alabama and to Texas. In sandy lowland streams, particularly in muddy backwaters.

EASTERN BANDED DARTER—*Etheostoma zonale zonale* (Cope). Fig. 216—From northwestern and central Iowa and from Minnesota and Wisconsin (Mississippi River and Lake Michigan drainages), through Illinois (including one record from the Lake Michigan basin), Indiana, Ohio (including one questioned record from the Lake Erie tributary system) to the Ohio River drainage area of New York and Pennsylvania; south to the Tennessee River system of North Carolina and Tennessee (this subspecies may also inhabit part of the range of the species in the Ozark Upland of Arkansas, southern Missouri and eastern Oklahoma). Typically on riffles of moderate-sized streams, particularly in algae and other vegetation.

Most and possibly all of the Ozark Upland should be listed for the habitat of *E. z. arcansanum* Jordan and Gilbert. The Chickasawha River in Mississippi and other central Gulf tributaries are the home of the form *elegans,* which is either a subspecies of *zonale* or a very closely related species.

NOTE.—Two species of darters that were retained on tenuous basis in the preceding edition of this book are now removed. One is the bluebreast darter, *Etheostoma camurum* (Cope) (Fig. 215), the single Great Lakes record of which, from the Lake Erie drainage of Ohio, is regarded by M. B. Trautman as definitely erroneous. The other is the mud darter, *Etheostoma asprigene* (Forbes) (Fig. 217), formerly listed as *Poecilichthys jessiae asprigenis* Forbes, the only Great Lakes record of which is now regarded as having probably been based on specimens of *E. exile.*

IOWA DARTER—*Etheostoma exile* (Girard). Fig. 218 and color plate no. 32.—From Montana and from Alberta through the James Bay drainage to the upper St. Lawrence basin in Quebec, and to Lake Champlain; to Quebec, and northern United States from Montana to Lake Champlain; southward to southern tributaries of Lake Ontario, to the Allegheny drainage of New York and to the northern parts of Ohio, Indiana and Illinois; and in the west to Iowa, Nebraska and Colorado (with one population, perhaps subspecifically distinct, in northern Arkansas). Abundant throughout the Great Lakes drainage area, on sandy to muddy bottoms about vegetation, in moderately cool lakes and sluggish streams.

RAINBOW DARTER—*Etheostoma caeruleum* Storer. Figs. 219 and 220.—From the Mississippi River drainage of Iowa, Minnesota and Wisconsin, and the Great Lakes basin of Michigan (south from the Pere Marquette, Manistee and Cheboygan systems), to the Lake Huron and Lake Erie tributaries of Ontario, to the western half of the Lake Ontario drainage in Ontario and New York, with some populations in the upper St. Lawrence and Ottawa river drainages of Quebec; southward in the Ohio and Mississippi valleys to the northern parts of Alabama and Arkansas (several of the southern races will probably be separated subspecifically, thus restricting the range of the typical subspecies). Ordinarily on gravel in creeks.

NORTHERN ORANGETHROAT DARTER—*Etheostoma spectabile spectabile* (Agassiz). Fig. 221.—From eastern Kansas through most of Missouri and parts of northern Arkansas, to Iowa, Illinois, southeastern Michigan and the western three-fifths of Ohio; southeastward (as distinct subspecies?) to the Tennessee River system of Tennessee and Virginia (generally rare south of the Cumberland and Duck drainages) replaced southwestward to eastern Texas by other subspecies. Frequenting the shallow, gravelly riffles in relatively slow to moderately swift waters of smaller creeks.

FANTAIL—*Etheostoma flabellare* Rafinesque.—The subspecies follow.

BARRED FANTAIL—*Etheostoma flabellare flabellare* Rafinesque. Figs. 222 and 223.—From Indiana and the southern two-thirds of the Lower Peninsula of Michigan (chiefly in the eastern drainage), through southern Ontario to the St. Lawrence–Lake Champlain basin and the Mohawk River system (*via* canal?) and the northern part of the Susquehanna River system; southward in the eastern part of the Ohio Valley to West Virginia, North Carolina and Tennessee (subspecies in the Middle Atlantic states not determined). Most often on the gravel bottom of the slower and shallower riffles in small streams; also in some lakes, including Great Lakes shores.

STRIPED FANTAIL—*Etheostoma flabellare lineolatum* (Agassiz). Col. pl. no. 33.—From Minnesota through the entire Mississippi River drainage of Wisconsin to the Lake Michigan drainage of that state and of the western two-thirds of the Upper Peninsula of Michigan; south through Illinois and Iowa to the western parts of the Ohio River system in Indiana (at least as intergrades with *E. f. flabellare*), Kentucky, Tennessee and northern Alabama; southwest to northern Arkansas and northeastern Oklahoma. Habitat as for *E. f. flabellare.*

LEAST DARTER—*Etheostoma microperca* Jordan and Gilbert. Figs. 224 and 225.—From the Mississippi River system in Minnesota eastward through the Michigan-Huron basin, the southern drainage of Lake Superior and the Lake Erie system to tributaries of the western part of Lake Ontario in Ontario; south to Kentucky, southeastern and southwestern Missouri, northwestern Arkansas, and southern Oklahoma (the southwestern population possibly representing a distinct subspecies). Preferring dense vegetation over soft bottom, in quiet waters of streams and in lakes.

NORTHERN GREENSIDE DARTER—*Etheostoma blennioides blennioides* Rafinesque. Figs. 45, 226, 227 and col. pl. no. 34.—Southeastern Michigan (eastern drainage except for one record each in the headwaters of the Grand River and the St. Joseph River systems), the Lake St. Clair drainage of Ontario and the southern part of the Lake Ontario basin; also (probably *via* canal route) in the Mohawk River system; south to the northern parts of the Ohio River system from eastern Illinois to western Pennsylvania and New York. Records from southeastern Wisconsin, Chicago River and Iowa are regarded as incorrect. Preferring riffles where the rocks are coated with green algae. The other subspecies inhabit the Tennessee and Ozark uplands.

SUNFISH FAMILY—*Centrarchidae*

(Figs. 228–242 and col. pls. nos. 35–41)

The sunfish family is a diverse group of spiny-rayed fishes which have the spinous and soft-rayed portions united into a single dorsal fin, rather than being separated into two distinct fins as in the perch family. The largemouth bass and, to a lesser degree, the smallmouth bass show an incomplete division of the dorsal fin into these two parts. Another spiny-rayed fish in this fauna, the freshwater sheepshead, has two parts of the dorsal fin continuous, but it is distinguished by the extension of the lateral line across the tail fin. Sunfish scales are weakly to moderately ctenoid, rarely cycloid. A single individual of some kinds may exhibit both scale types.

Members of this exclusively North American freshwater family fall into the following groups: (1) largemouth and smallmouth basses (the common black basses of anglers); (2) the crappies (and the round sunfish, *Centrarchus,* which is not in the Great Lakes drainage); (3) the true sunfishes and rock bass (and the Sacramento perch, *Archoplites,* of the Pacific Coast). The number of recognized species in the three groups is 4, 3 and 18 respectively.

In the Great Lakes region these fishes inhabit all kinds of waters. They are all spring spawners and utilize shallow depressions excavated by the males for nests. Spawning for some kinds extends into the summer. The males guard the eggs and the larvae, and in some species herd the young. Like other essentially carnivorous fish, the smallest young first feed on microscopic organisms, and later, upon aquatic invertebrates. Larger individuals, particularly of the basses and crappies, are chiefly piscivorous. Rate of growth and sizes attained are definitely related to amount of food available. Sunfishes in particular become stunted when their populations are too large for the food supply. As a result the size in many waters is too small for satisfactory angling. The tendency of members of this family, particularly the sunfishes, to overpopulate waters in spite of angling pressure is currently bringing about liberalization of fishing laws and an abandonment of artificial propagation and stocking. The usual sizes taken by fishermen vary for the different species. Examples for common kinds in the Great Lakes region are: basses, mostly between 12 and 16 inches; crappies, mostly 8 to 9 inches but often exceeding 12 inches; bluegills and rock bass, mostly 7 to 8 inches but may go to more than 12 inches; green sunfish, rarely 6 inches; longear sunfish, seldom 5 inches.

Angling methods for members of this family are various. For the basses bait casting, fly casting, "bass bugging" and the use of the most common live baits are all successful. Smallmouth bass are particularly vulnerable to fishing after dark, by casting the lure into the shallows near reeds or other cover and retrieving it slowly. As a rule, either in open water or through the ice, live minnows are the most effective bait for crappies. The sunfishes and rock bass are readily taken with wet and dry flies or with earthworms. Sunfishes may be caught through the ice with grubs or "wigglers" (burrowing mayfly larvae) or with artificial "ice flies."

1. Anal spines 3 (very rarely 2 or 4); dorsal spines usually 10 2
   Anal spines usually 6 (5 to 7); dorsal spines not normally 10 (either 6 to 8 or 11 to 14, rarely 10) 10

2. Scales small, 58 or more in lateral line; body depth about one-third standard length; precaudal vertebrae typically 15 (black basses) .. 3
   Scales large, 55 or fewer in lateral line; body depth usually about one-half standard length; precaudal vertebrae typically 12 4

3
: Dorsal fin shallowly emarginate, the shortest spine at the emargination being more than one-half of the longest (Fig. 16); anal and soft dorsal fins with their bases bearing small scales on the membranes; upper jaw extending beyond middle of pupil but not to hind margin of eye (mouth closed); 68 to 81 scales in lateral line; cheek scales in 14 to 18 rows; pyloric caeca typically unbranched at their bases.
: NORTHERN SMALLMOUTH BASS—*Micropterus dolomieui dolomieui* Lacépède. (Figs. 228, 229 and col. pl. no. 35)

: Dorsal fin almost divided, the shortest spine at the emargination less than half as long as longest (Fig. 17); anal and soft dorsal fins without scales on membranes at their bases; upper jaw extending beyond hind margin of eye in adult (mouth closed); 58 to 69 scales in lateral line; cheek scales in 9 to 12 rows; pyloric caeca typically branched at their bases.
: NORTHERN LARGEMOUTH BASS—*Micropterus salmoides salmoides* (Lacépède). (Figs. 230, 231 and col. pl. no. 36)

4
: Teeth present on tongue, ectopterygoid, and entopterygoid; supramaxilla well developed, its length greater than the greatest width of the maxilla (Fig. 47).
: WARMOUTH—*Chaenobryttus gulosus* (Cuvier). (Fig. 232)

: No teeth on tongue, ectopterygoid, or entopterygoid; supramaxilla reduced or absent, its length when present less than greatest width of maxilla (*Lepomis*) 5

Fig. 47.

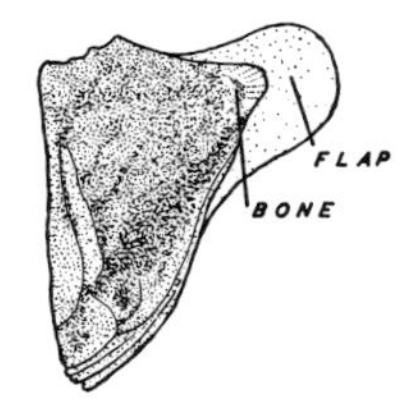

Fig. 48.

Fig. 47. Supramaxilla well developed, its length greater than the greatest width of the maxilla, the bone immediately below the supramaxilla (warmouth, *Chaenobryttus gulosus*).

Fig. 48. Opercular bone extending only to middle of "ear flap." In this species (orangespotted sunfish, *Lepomis humilis*) the tip of the bone is so flexible as to be distinguishable from the membranous flap only by close inspection.

5
: Pectoral fins short and rounded, contained about 4 times in standard length 6

: Pectoral fins long and pointed, longest rays, near dorsal side of fin, contained a little less to a little more than 3 times in standard length .. 7

6
: Scales smaller, 44 or more in lateral line; opercle stiff to margin, its membrane wide and light-colored; gill-rakers long and slender (Fig. 49B).
: GREEN SUNFISH—*Lepomis cyanellus* Rafinesque. (Figs. 49B, 233 and col. pl. no. 37)

: Scales larger, 39 or fewer in lateral line; opercle flexible posteriorly, its membrane with a narrow red margin behind and usually below; gill-rakers reduced to knobs (Fig. 49D).
: NORTHERN LONGEAR SUNFISH—*Lepomis megalotis peltastes* Cope. (Figs. 49D, 239)

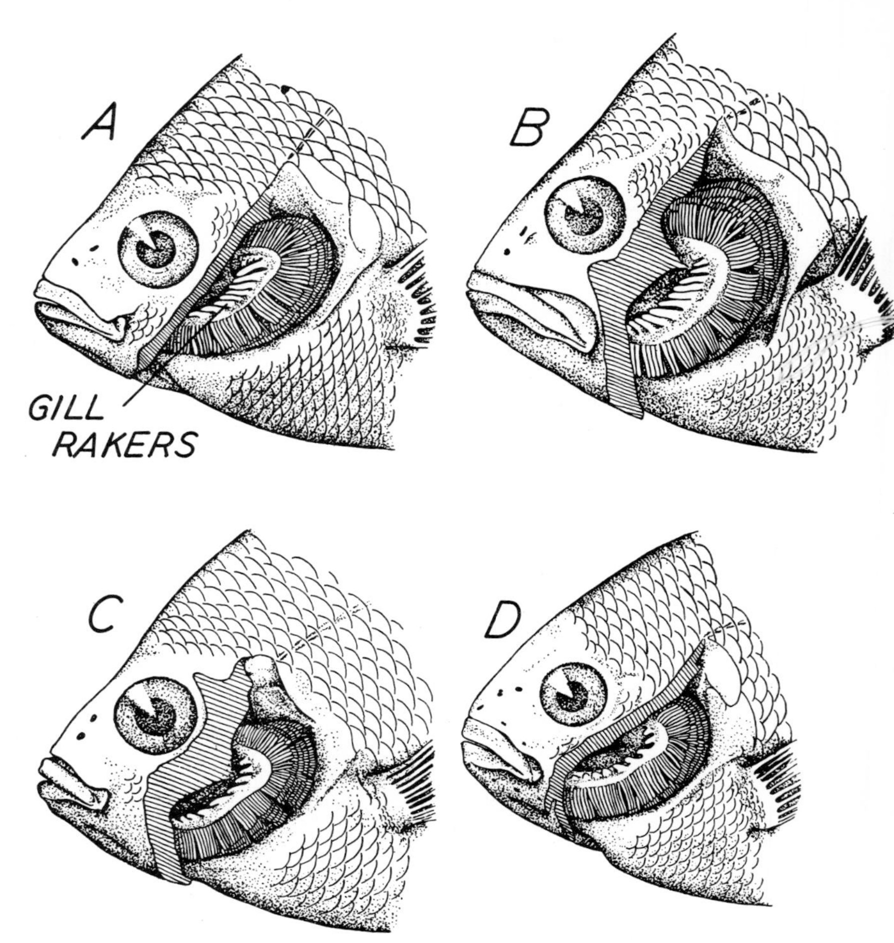

Fig. 49. Heads of four sunfishes for comparison of mouth size and gill-raker proportions. A. Common bluegill (*Lepomis m. macrochirus*). B. Green sunfish (*L. cyanellus*). C. Pumpkinseed (*L. gibbosus*). D. Northern longear sunfish (*L. megalotis peltastes*).

7
- Opercle (the bone, not the membrane; see Fig. 48) stiff behind and never fimbriate or ragged at its caudal margin; gill-rakers short and stout (as in Fig. 49C) 8
- Opercular bone flexible posteriorly and sometimes quite fimbriate or ragged at its caudal margin; gill-rakers long and slender (as in Fig. 49A) 9

8
- Opercle with a definite scarlet spot; cheeks with prominent blue and orange stripes in life; pectoral fins less than one-third standard length in adult; body outlines gibbous.
  PUMPKINSEED—*Lepomis gibbosus* (Linnaeus). (Figs. 49C, 234, 235)
- Opercle with a broad scarlet margin; cheeks without conspicuous orange and blue streaks; pectoral fins more than one-third standard length in adult; form more rhomboidal.
  WESTERN REDEAR SUNFISH—*Lepomis microlophus* (Günther), subspecies.

9 Opercular bone extending almost to margin of "ear flap"; sensory cavities of head small; anal soft-rays typically 10 to 12.
COMMON BLUEGILL—*Lepomis macrochirus macrochirus* Rafinesque. (Figs. 49A, 236 and col. pl. no. 38)

Opercular bone extending only to middle of "earflap" (Fig. 48); sensory cavities of head very large; anal soft-rays 7 to 9.
ORANGESPOTTED SUNFISH—*Lepomis humilis* (Girard). (Figs. 237, 238)

10 Dorsal spines 11 or 12; anal fin base about one-half length of dorsal base; branchiostegal rays 6; margin of preopercle entire or nearly so.
NORTHERN ROCK BASS—*Ambloplites rupestris rupestris* (Rafinesque). (Fig. 240 and col. pl. no. 39)

Dorsal spines 6 to 8; anal fin base about equal to length of dorsal base; branchiostegal rays 7; margin of preopercle finely toothed or serrate (*Pomoxis*) 11

11 Dorsal spines normally 7 or 8; length of dorsal fin base about equal to distance from origin of dorsal to eye; caudal vertebrae typically 19; body more speckled.
BLACK CRAPPIE—*Pomoxis nigromaculatus* (LeSueur). (Fig. 242 and col. pl. no. 41)

Dorsal spines normally 6; length of dorsal fin base much less than distance from origin of dorsal to eye; caudal vertebrae typically 18; body more definitely banded.
WHITE CRAPPIE—*Pomoxis annularis* Rafinesque. (Fig. 241 and col. pl. no. 40)

NORTHERN SMALLMOUTH BASS—*Micropterus dolomieui dolomieui* Lacépède. Figs. 228, 229 and color plate no. 35.—Originally from northern Minnesota to Lake Nipissing and Quebec (with a relict population in the Nipigon River and south to the Tennessee River drainage of Alabama and to eastern Oklahoma (as intergrades with *M. d. velox* Hubbs and Bailey, which occupies tributaries of the Arkansas River in northeastern Oklahoma, western Arkansas, and Missouri). Introduced extensively elsewhere. Associated with clear, cool rivers, and lakes; in streams preferring those of medium size, with pools and riffles.

NORTHERN LARGEMOUTH BASS—*Micropterus salmoides salmoides* (Lacépède). Figs. 230, 231 and color plate no. 36.—Originally from southern Canada (Ontario northward to Lake Temiskaming and Quebec) throughout the entire Great Lakes system and Mississippi Valley to northeastern Mexico (subspecies uncertain) and extreme western Florida, and north along the Coastal Plain to North Carolina and possibly farther (intergrading from western Florida to southwestern South Carolina with *M. s. floridanus* (LeSueur) of peninsular Florida) introduced elsewhere extensively. Typically in weedy or brushy, mud-bottomed lakes and ponds, bayous, backwaters and sluggish streams.

WARMOUTH—*Chaenobryttus gulosus* (Cuvier). Figs. 47 and 232.—From Kansas and Iowa to the Mississippi River drainage in Wisconsin, the southern two-thirds of the Lower Peninsula of Michigan, Lake Erie and the Allegheny River tributaries of Pennsylvania; southward to the Rio Grande and to Florida, and northward in the Atlantic drainage to Pennsylvania and southern New York (northern limit as native uncertain). Usually in sluggish, very weedy waters, over a soft bottom.

GREEN SUNFISH—*Lepomis cyanellus* Rafinesque. Figs. 49B, 233 and color plate no. 37.—From Colorado and South Dakota through Minnesota, Wisconsin and the Upper Peninsula of Michigan (rare) to extreme southern

Ontario (where also rare) and western New York; southward through the lower Great Lakes and Mississippi basins to Georgia and southern Alabama, northeastern Mexico and New Mexico. Introduced elsewhere, and the stated limits of its range may in part have resulted from plantings. A species of lakes, ponds and pools of creeks; often commonest in sluggish creeks and small ponds.

PUMPKINSEED—*Lepomis gibbosus* (Linnaeus). Figs. 49C, 234 and 235.—Southern Canada, and from the Dakotas and the Red River of the North to the Maritime Provinces, south along the Atlantic coast to the Savannah River system between South Carolina and Georgia, and in the Mississippi River system to western Pennsylvania, Ohio, and Iowa. Rather widely introduced elsewhere. Common throughout the Great Lakes region. Preferring weedy lakes and ponds, and similar parts of streams; in cool to moderately warm waters.

WESTERN REDEAR SUNFISH—*Lepomis microlophus* (Günther), subspecies.—From the Mississippi River in Missouri (and, formerly, in Iowa) and from southern Indiana (and, presumably through recent introductions, in the Lake Michigan drainage of Indiana and in southern Michigan), south to Alabama and the Rio Grande. Now being introduced elsewhere. Represented in Alăbama, Georgia, and Florida by the typical subspecies, *L. m. microlophus* (Günther). Commonest in and near large, warm rivers, bayous and lakes.

COMMON BLUEGILL—*Lepomis macrochirus macrochirus* Rafinesque. Figs. 49A, 236 and color plate no. 38.—From Minnesota through southern Ontario to the adjacent upper St. Lawrence drainage and to Lake Champlain; southward through the Great Lakes and Mississippi River basins to the Chattahoochee River in Georgia and westward, intergrading in the Arkansas and Red river systems with *L. m. speciosus* (Baird and Girard), which ranges through Texas to northeastern Mexico. Introduced elsewhere. Most commonly in lakes and ponds having a moderate amount of vegetation; in streams generally restricted to the quieter pools.

The distribution just given covers the full range of the species, except for the Atlantic Coast streams from the Carolinas to Florida, which are occupied by *L. m. purpurascens* Cope (possibly a distinct species).

ORANGESPOTTED SUNFISH—*Lepomis humilis* (Girard). Figs. 48, 237, and 238.—From the James River in North Dakota through the Mississippi drainage of southern Minnesota, southern Wisconsin, Illinois and Indiana to western Ohio (in Lake St. Marys, which flows into Lake Erie as well as into the Wabash River, and in all of the Maumee River basin) and to extreme southeastern Michigan; south, through the western parts of Kentucky and Tennessee, to the Tennessee River system in Alabama and to western Mississippi; through the Plains region to Texas, western Oklahoma, and Colorado. Inhabiting various types and sizes of streams and lakes; commonly in silty water. With the modification of streams as a result of farming this species has been spreading its range northeastward.

NORTHERN LONGEAR SUNFISH—*Lepomis megalotis peltastes* Cope. Figs. 49D and 239.—Glacial lake regions from the northern parts of Iowa, Minnesota and Illinois, and from eastern Wisconsin and the entire Lower Peninsula of Michigan to the Lake Erie and the Lake Huron drainages of Ontario, the Lake Erie tributaries of Ohio, the southern part of the Lake Ontario basin, the St. Lawrence drainage in Quebec and parts of the Allegheny River system of Pennsylvania. Chiefly in the more sluggish waters of clear lakes, ponds and streams; often associated with a moderate amount of vegetation.

The other subspecies of *L. megalotis* extend the range of the species to the Gulf states and to the Rio Grande tributaries of Mexico.

NORTHERN ROCK BASS—*Ambloplites rupestris rupestris* (Rafinesque). Fig. 240 and color plate no. 39.—From southern Manitoba and the Red River drainage of North Dakota through southern Ontario (English River and

Lake Abitibi southward, but not Lake Nipigon) and the Great Lakes and the St. Lawrence River system to Quebec and Lake Champlain; southward, west of the Appalachian Divide, to the Tennessee River drainage of Alabama and North Carolina and to the Ozark region in northern Arkansas and eastern Oklahoma. Intergrading in southeastern Missouri and eastern Arkansas with the Gulf coast form *A. r. ariommus* Viosca, which ranges from western Florida to Texas. Now common on the Atlantic slope, presumably as the result of movements through canals and through introductions. Introduced elsewhere. Usually restricted to cool, weedy lakes and rocky streams.

WHITE CRAPPIE—*Pomoxis annularis* Rafinesque. Fig. 241 and color plate no. 40.—From Nebraska and from Big Stone Lake, Minnesota, through western and southeastern Wisconsin and the southern two-thirds of the Lower Peninsula of Michigan to southern Lake Huron and the Lake Erie basin in Ontario and the northern parts of the Lake Ontario drainage (and possibly the St. Lawrence River); south through the Mississippi Valley to Texas and southern Alabama, and north along the Atlantic slope to North Carolina. Stocked in the West. Generally rare in the Great Lakes. Most often in turbid rivers, sloughs and lakes; not averse to a mud bottom.

BLACK CRAPPIE—*Pomoxis nigromaculatus* (LeSueur). Fig. 242 and color plate no. 41.—Southern Canada from Manitoba and Lake of the Woods to the Ottawa River and to the upper St. Lawrence River in Quebec, south to Lake Champlain and through the Great Lakes; in the Mississippi River drainage from eastern Nebraska and from Minnesota to western New York and Pennsylvania, south to northern Texas and northern Florida; thence northward near the coast to North Carolina (introduced farther north, and also in the West, northward to British Columbia). In the Great Lakes region chiefly in rather clear, weedy lakes; southward usually in larger streams, lakes, and impoundments; not so common as *P. annularis* in silty waters.

SILVERSIDE FAMILY—*Atherinidae*

(Fig. 243 and col. pl. no. 42)

These are streamlined fishes, more or less transparent in life, with a small spinous first dorsal fin which may escape notice unless carefully sought. The anal fin is also diagnostic since it has so many rays. The scales are very thin and deciduous and are cycloid.

Most members of this family are marine fishes, but many species live in fresh waters. The family is widespread in tropical and temperate regions. All but three of the New World species are referable to an exclusively American subfamily, the Atherinopsinae. The single species of the Great Lakes region, the brook silverside, belongs to this subfamily.

The brook silverside, a surface swimmer, is common in lakes and in the quieter parts of streams. It often skips into the air for a short distance. A common local name resulting from this habit is "skipjack." The breeding season is in the late spring. The eggs are unique in that each has a sticky thread-like process that serves both as a flotation organ and subsequently as a holdfast when the egg finally comes in contact with an object.

Growth is rapid; the young attain the adult size of a few inches and spawn at the age of one year. The usual life span is less than two years.

Brook silversides are eaten by predacious fishes including game kinds. They can be used as bait minnows but they are tender and do not hold up well.

NORTHERN BROOK SILVERSIDE—*Labidesthes sicculus sicculus* (Cope). Fig. 243 and color plate no. 42.—From Minnesota, the western and eastern drainages of Wisconsin and the Lower Peninsula of Michigan, through the Georgian Bay–Lake Erie portions of Ontario to the upper St. Lawrence River system and to the Allegheny River drainage of New York and Pennsylvania; also in the Mohawk River system of New York (presumably as a canal immigrant); southward to the Gulf states, Texas and Oklahoma. Represented to the southeast by *L. s. vanhyningi* (Bean and Reid); southwestern form possibly also distinct. Characteristic of surface waters in lakes and in lake-like habitats of streams.

DRUM FAMILY—*Sciaenidae*

(Fig. 244 and col. pl. no. 43)

The drums are distinguishable from all other bass-like spiny-rayed fishes by the extension of the lateral line across the tail fin. They make a purring, characteristic sound. The scales are ctenoid.

With a few exceptions, including the species here treated, the drums are all marine shore fishes. There are scores of species in tropical and temperate waters. The group is particularly well represented in Panama. Many of the species are important food fishes, and some, like the white seabass (*Cynoscion nobilis*) of California, are highly prized by deepsea anglers.

The freshwater drum, or "sheepshead" is here essentially a fish of the shallow waters of the large lakes themselves, excepting Lake Superior. It is a bottom dweller and seems to prefer turbid situations to clear ones. This drum is a spring spawner, but little is known about the reproductive habits. The smallest individuals feed upon microscopic organisms. At lengths of a few inches, they consume chiefly aquatic insects and small crustaceans. With increase in size, they eat more and more snails and clams, which they crush with the molar teeth in the throat.

The freshwater drum is not a food fish of high merit, although it enters the commercial catch and is taken occasionally by anglers. Most of the commercial take is from Lake Erie. The average size netted in Lake Erie is about fifteen inches and a pound and a half. Larger sizes, to ten pounds, are reported, and in early years weights approaching 100 pounds were attained. The size of the pharyngeal (throat) bones and molar teeth that are found in the camp sites of Indians indicate that in primordial times sizes of about 200 pounds were attained. The Indians apparently regarded these bones, as well as the lucky bones of the ear, as of special value or significance, for large numbers of the remains have been found about old Indian settlements—some far from the range of the fish.

FRESHWATER DRUM—*Aplodinotus grunniens* Rafinesque. Fig. 244 and color plate no. 43.—From the Hudson Bay drainage of southwestern Saskatchewan and southern Manitoba and east-central Ontario (northward to Lake Abitibi) and from the Great Lakes (other than Superior) to Quebec and Lake Champlain; in the Mississippi lowlands from Montana and Nebraska to Pennsylvania and south to the Gulf states, and through eastern Mexico to the Río Usumacinta system in Guatemala. Generally in large rivers and lakes, usually in silty waters.

SCULPIN FAMILY—*Cottidae*

(Figs. 245–248)

Most sculpins, and all members of this family in the Great Lakes fauna, are characterized by enlarged, flattened heads and by expansive pectoral fins. The preopercular spine on the side of the head is a feature. They are spiny-rayed, but the spines are so feeble as to require careful examination to be appreciated (*e. g.*, Fig. 50). Scales are lacking entirely or are represented by dermal prickles, commonly only behind the pectoral fin bases.

The family Cottidae is essentially marine. Only the one large genus *Cottus* is exclusively freshwater. *Myoxocephalus* occurs in the Arctic Ocean and the boreal parts of the other northern oceans as well as in the relict waters of the Ice Age. There are scores of species in the North Pacific, smaller numbers in the Arctic Ocean and in the North Atlantic and a very few in the Antarctic.

The "muddlers," as these fishes are sometimes known in the Great Lakes region, are bottom dwellers. Various species are found in streams and in the Great Lakes, both about the shores and at considerable depths. The habits of the fourhorned sculpin are poorly known; it is, however, an important food of the lake trout and burbot. The northern mottled sculpin is principally a stream fish; the slimy species lives in colder trout streams and about the Great Lakes shores; and the Great Lakes mottled kind is essentially a shore-zone inhabitant.

The following comments apply mostly to the three species last named above. In any of the major habitats, the principal situations occupied are those with a rocky bottom, where the sculpins retreat under stones during daylight hours. In the spring the eggs are attached in clusters, usually to the undersides of stones. The male drives the female into the nest and then guards the eggs. Stomach contents include small amounts of algae and aquatic insects and their larvae. The larger individuals eat fish (sometimes including smaller sculpins). If fish eggs are available they may be eaten. Trout-stream sculpins are accused of being predatory on trout eggs and young, but there is only a little evidence, which is both postive and negative, on this score. Probably most of the trout eggs consumed are loose ones that fail to become buried in the nests. Sculpins are sometimes used as bait by anglers.

1 { Gill-membranes free from isthmus; second preopercular spine conspicuous, directed backward; dorsal fins widely separated.
GREAT LAKES FOURHORNED SCULPIN—*Myoxocephalus quadricornis thompsonii* (Girard). (Fig. 245)

Gill-membranes attached to side of wide isthmus far behind eye; second preopercular spine concealed by skin, directed downward; dorsal fins scarcely separated (*Cottus*) 2

2 { Preopercular spine long and spirally curved; head spatulate in outline when viewed from above; lateral line complete; body sometimes wholly prickly.
SPOONHEAD SCULPIN—*Cottus ricei* (Nelson). (Fig. 246)

Preopercular spine short and little curved; head broadly rounded, semi-oval as seen from above; lateral line normally terminating below base of second dorsal; body never wholly prickly 3

3 { Pelvic rays typically I, 3* or the fourth soft-ray variously reduced on one or both sides; palatine teeth (see Fig. 27 for position in roof of mouth) typically not developed; skin slimier in life.
EASTERN SLIMY SCULPIN—*Cottus cognatus gracilis* Heckel. (Fig. 248)

Pelvic rays typically I, 4 (Fig. 50)*; palatine teeth normally present; skin less slimy (*Cottus bairdii*) 4

4 { Distance between tip of snout and anus when measured backward from anus extending to a point nearer base than end of caudal (except in some very large specimens; to near base of caudal in young); body usually rather robust; dark bars definitely developed.
NORTHERN MOTTLED SCULPIN—*Cottus bairdii bairdii* Girard. (Fig. 247)

Distance between tip of snout and anus when measured backward from anus extending to a point nearer end than base of caudal (to near middle of caudal in young); body averaging more slender; dark bars less distinctly developed.
GREAT LAKES MOTTLED SCULPIN—*Cottus bairdii kumlieni* (Hoy).

* The spine is a slender splint bound in the membrane of the first soft-ray and not discernible without dissection. It is always present. (See Fig. 50.)

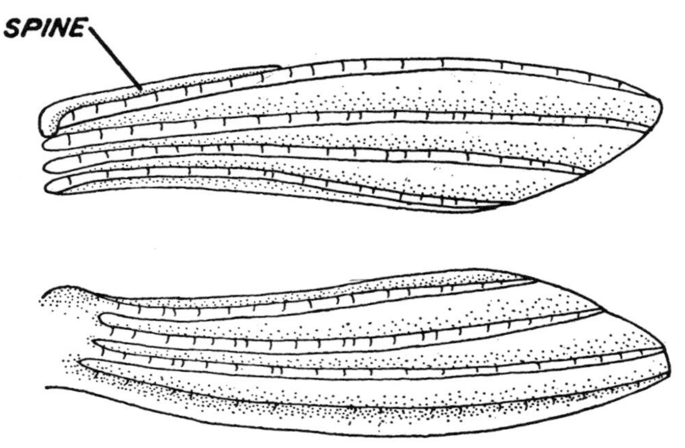

Fig. 50. Ventral view of pelvic fins of northern mottled sculpin, *Cottus b. bairdii,* with radial formula, I, 4. The fleshy sheath which surrounds the spine and first ray, making them appear deceptively as one element, has been removed from the left fin (upper figure). The spine is always present.

GREAT LAKES FOURHORNED SCULPIN—*Myoxocephalus quadricornis thompsonii* (Girard). Fig. 245.—Streams of Arctic Canada and the deep waters of all the Great Lakes, Lake Nipigon and Torch Lake, Michigan. The species occurs in the Polar Basin, Bering Sea, and North Atlantic Ocean, with glacial relicts in northern lakes and streams of Eurasia and North America.

SPOONHEAD SCULPIN—*Cottus ricei* (Nelson). Fig. 246.—Canada from the delta of the Mackenzie River and from its tributaries in British Columbia, across the Mackenzie and Saskatchewan river drainages to Hudson Bay and tributary waters east as far as Lake Abitibi; in all of the Great Lakes proper, descending to deep water; recorded from inland lakes in the Great Lakes basin only on Isle Royale and from Lake Charlevoix, Michigan.

MOTTLED SCULPIN—*Cottus bairdii* Girard.—The range of the species extends as far as the Alabama River system of Georgia, which is occupied by *C. b. zopherus* (Jordan), and a few locations in Missouri. The northern limits are indicated below.

NORTHERN MOTTLED SCULPIN—*Cottus bairdii bairdii* Girard.—From the southern parts of eastern Canada (including several Hudson Bay tributaries and Labrador), and from Minnesota through Wisconsin to northern Illinois and Indiana; throughout the Great Lakes region (including Lake Erie, at least its western end, but excluding the other Great Lakes and the inland lakes inhabited by *C. b. kumlieni*) to Quebec and New Brunswick; common in northern Ohio, with relict colonies in southern Ohio; in some mountain streams of the Ohio River system in Pennsylvania, West Virginia, Kentucky, Tennessee, and Alabama (at higher elevations southward); in the Atlantic drainage of the United States confined to certain upper tributaries of the Susquehanna, Potomac and James watersheds, from New York to Virginia. Represented southeastward and perhaps in the Ozarks by other subspecies. In cool creeks and in many small northern lakes.

GREAT LAKES MOTTLED SCULPIN—*Cottus bairdii kumlieni* (Hoy).—Entire margins of lakes Superior, Michigan, Huron and Ontario and the eastern basin of Lake Erie, the Finger Lakes and the St. Lawrence River, scarcely ascending streams; also in a very few inlands lakes of Michigan where conditions approach those of the Great Lakes, and Lake Attawapiskat, James Bay drainage.

EASTERN SLIMY SCULPIN—*Cottus cognatus gracilis* Heckel. Fig. 248.—From the Great Lakes and probably from the southern part of the Arctic drainage of central Canada northeast to Ungava Bay and the Labrador Peninsula; south on the Atlantic slope (common northward and scatteringly southward) to the Potomac watershed in West Virginia and the James River system in Virginia; and to the basins of all of the Great Lakes (one record, in need of verification, from the Allegheny drainage of New York); in the west to northern Minnesota and, in isolated relict colonies, to northeastern Iowa. In all the Great Lakes, including the deeper, eastern part of Lake Erie; occurring inland through the entire Lake Superior basin, in Michigan south to the White and Rifle river systems, and in the Finger Lakes and their tributaries in New York. Represented in the northern part of the Arctic drainage of Canada, westward to British Columbia, by *C. c. cognatus* Richardson. That subspecies abounds in Great Slave Lake and Great Bear Lake, in Lake Bennett and Teslin Lake (B. C.) of the Yukon system and in the Nushagak and Copper rivers of Alaska, and is widespread in central and Arctic Alaska. The species ranges southward as far as the Columbia River system in eastern Washington, and may occur in eastern Siberia.

STICKLEBACK FAMILY—*Gasterosteidae*

(Figs. 249–251 and col. pl. no. 44)

The sticklebacks are spiny-rayed fishes in which dorsal spines are separated from one another and each is connected to the back with its own fin membrane. Furthermore the caudal peduncle is slenderer than in other Great Lakes fishes and the pelvic fins are each reduced to a single stout spine with no more than rudiments of one or two soft-rays.

About a dozen species of sticklebacks occur in northern seas and fresh waters. They occur along Arctic shores and on both sides of the North Atlantic and of the North Pacific.

The threespine and ninespine sticklebacks are principally shore inhabitants of the larger lakes. The brook stickleback thrives both in streams of various sizes and in bog ponds. In the spring sticklebacks construct an elaborate nest somewhat resembling that of an oriole. Fewer than a hundred eggs are usually produced and these are zealously guarded by the male. The sticklebacks have little direct economic significance. In many waters, however, they are effective in the control of mosquitoes, for they feed on the "wrigglers" and often live in habitats that other fishes can not tolerate.

1. Gill-membranes united medially to isthmus, not forming a free fold; lateral plates developed at least in part; dorsal spines 3.
   THREESPINE STICKLEBACK—*Gasterosteus aculeatus* Linnaeus. (Fig. 251)

   Gill-membranes confluent, forming a broad free fold across isthmus; lateral plates wholly absent; dorsal spines 5 to 11 2

2. Dorsal spines 5 or 6, scarcely divergent; tail without trace of keel; caudal peduncle deeper than wide; caudal fin rounded.
   BROOK STICKLEBACK—*Eucalia inconstans* (Kirtland). (Fig. 249 and col. pl. no. 44)

   Dorsal spines 8 to 11, strongly divergent; tail with a sharp lateral keel; caudal peduncle much wider than deep; caudal fin lunate.
   NINESPINE STICKLEBACK—*Pungitius pungitius* (Linnaeus). (Fig. 250)

BROOK STICKLEBACK—*Eucalia inconstans* (Kirtland). Fig. 249 and color plate no. 44.—In Canada from the Peace River system of northern British Columbia and from Alberta to Hudson and James bays and to Quebec and New Brunswick (doubtfully from Greenland); in the United States from Montana to the St. Lawrence, Champlain, Hudson, Susquehanna and Allegheny drainages of New York, and to northern Maine; south to the systems of the Ohio River in Pennsylvania, Ohio and Indiana, the Illinois River in Illinois and the Missouri River in Kansas. Throughout the Great Lakes basin. Never marine; common in boggy lakes and streams, alkali lakes, spring holes and many trout streams, and in the Great Lakes taken as deep as 17 fathoms.

NINESPINE STICKLEBACK—*Pungitius pungitius* (Linnaeus). Fig. 250.—Circumpolar in Europe, Asia, and North America; in virtually all Arctic drainages of Eurasia and North America but absent from Greenland and Iceland; south in the Pacific region to Kamchatka, Mongolia, southeastern Siberia, China and Japan, and to the Alaska Peninsula in the east; south in the Atlantic to New Jersey in the west and to central Europe in the east; northern limit not accurately known (from Walters, 1955, who has also recognized two subspecies). In the Great Lakes area almost wholly confined to the marginal waters of all the Great Lakes, to depths as great as 60 fathoms, including Lake St. Clair but excluding Lake Erie; south of Isle Royale and the Keweenaw Peninsula known from only four interior lakes in Michigan (Burt, Higgins, Gull and Birch), and from the Rapid River, Michigan, and Canandaigua Lake, New York. A fish of cool, quiet waters, salt and fresh.

THREESPINE STICKLEBACK—*Gasterosteus aculeatus* Linnaeus. Fig. 251.—Salt and fresh waters of the Northern Hemisphere, south to northern Africa, northern China, southern Japan, Baja California, the Hudson Bay region, and Chesapeake Bay. In the Great Lakes occurring only in the Lake Ontario basin (as intergrades between the circumarctic subspecies *aculeatus* and the eastern American, partially armored form, *cuvieri*).

---

*Footnote for List of References, p. 121.
The references listed here are, for the most part, those that have provided information on the taxonomy, range, and habitat of the fishes of the region. The student wishing additional bibliographic material is urged to consult the card catalog maintained by The Ontario Fisheries Research Laboratory, University of Toronto, and John Van Oosten's "Great Lakes Fauna, Flora and Their Environment, A Bibliography," published in 1957 by The Great Lakes Commission, University of Michigan, Ann Arbor.

LIST OF REFERENCES*

ADAMS, CHARLES C., and HANKINSON, THOMAS L.
1928. The Ecology and Economics of Oneida Lake. Fish. Bull. N. Y. St. Coll. Forestry, Syracuse Univ., Vol. 1, No. 4a (Roosevelt Wild Life Annals, Vol. 1, Nos. 3 and 4), pp. 235-548, figs. 175-244, 4 pls., map 16.

AITKIN, WALTER W.
1936. Some Common Iowa Fishes. Iowa State Coll. Extension Circ. 224, 33 pp., 27 figs.

ALLIN, A. E.
1951. The Fallfish, *Semotilus corporalis*, from the Lake Superior Drainage of Western Ontario. Copeia, 1951, No. 4, p. 300.

ALVAREZ, JOSE
1950. Claves para la Determinacion de Especies en los Peces de las Aguas Continentales Mexicanas. Mex. Secretaria de Marina, 144 pp., 16 figs.

AMERICAN FISHERIES SOCIETY
1948. A List of Common and Scientific Names of the Better Known Fishes of the United States and Canada. Amer. Fish. Soc., Ann Arbor, 45 pp.

ATTON, F. M., and JOHNSON, R. P.
1955. First Records of Eight Species of Fishes in Saskatchewan. Canad. Field Nat., Vol. 69, No. 3, pp. 82-84, 1 fig.

BACKUS, RICHARD H.
1951. New and Rare Records of Fishes from Labrador. Copeia, 1951, No. 4, pp. 288-294.
1957. The Fishes of Labrador. Bull. Amer. Mus. Nat. Hist., Vol. 113, Art. 4, pp. 275-337.

BAILEY, JOSEPH R., and OLIVER, JAMES A.
1939. The Fishes of the Connecticut Watershed. In: Biological Survey of the Connecticut Watershed. N. H. Fish and Game Dept., Surv. Rept. No. 4, pp. 150-189, figs. 57-80.

BAILEY, REEVE M.
1938. The Fishes of the Merrimack Watershed. In: Biological Survey of the Merrimack Watershed. N. H. Fish and Game Dept., Surv. Rept. No. 3, pp. 149-185, 12 text-figs.
1951. A Check-List of the Fishes of Iowa, with Keys for Identification. In: Iowa Fish and Fishing, Iowa State Cons. Comm., pp. 187-238, 9 figs., many text-figs.
1954. Distribution of the American cyprinid fish, *Hybognathus hankinsoni*, with comments on its original description. Copeia, 1954, No. 4, pp. 289-291, 1 fig.
1956. A Revised List of the Fishes of Iowa with Keys for Identification. In: Iowa Fish and Fishing, Third Ed., by James R. Harlan and Everett B. Speaker. Des Moines, Iowa Cons. Comm., pp. 327-377.

BAILEY, REEVE M., and HARRISON, HARRY M., JR.
1945. The Fishes of Clear Lake, Iowa. Iowa State Coll. Jour. Sci., Vol. 20, No. 1, pp. 57-77, 1 fig.

BAILEY, REEVE M., WINN, HOWARD ELLIOTT, and SMITH, C. LAVETT
1954. Fishes from the Escambia River, Alabama and Florida, with Ecologic and Taxonomic Notes. Acad. Nat. Sci. Philadelphia, Vol. 106, pp. 109-164, 1 fig.

BEAN, TARLETON H.
1903. Catalogue of the Fishes of New York. N. Y. St. Mus. Bull. 60, 784 pp.

BECKMAN, WILLIAM C.
1952. Guide to the Fishes of Colorado. Univ. Colo. Mus. Leaflet, No. 11, 110 pp., 14 figs., 106 text-figs.

BERG, LEO S.
1931. A Review of the Lampreys of the Northern Hemisphere. Ann. Mus. Zool. Acad. Sci. U. R. S. S., Vol. 32, pp. 87-116, 8 pls.
1940. Classification of Fishes, Both Recent and Fossil. Trav. Inst. Zool. Acad. Sci. U. R. S. S., Tome 5, Livr. 2, pp. 87-517, 190 figs. [Both Russian and English text.]

BERNER, LESTER M.
1948. The Intestinal Convolutions: New Generic Characters for the Separation of *Carpiodes* and *Ictiobus*. Copeia, 1948, No. 2, pp. 140-141, 4 figs.

BOND, CARL E., and BISBEE, LAWRENCE
1955. Records of the Tadpole Madtom, *Schilbeodes mollis*, and the Black Bullhead, *Ameiurus melas*, from Oregon and Idaho. Copeia, 1955, No. 1, p. 56.

BREDER, CHARLES M., JR.
1926. Locomotion of Fishes. Zoologica, Vol. 4, No. 5, pp. 159-297, figs. 39-83.

BRIDGE, THOMAS W., and BOULENGER, GEORGE A.
1904. Fishes. In: Cambridge Nat. Hist., Vol. 7, pp. 139-760, figs. 91-440.

BROWN, C. J. D.
1951. The Paddlefish in Fort Peck Reservoir, Montana. Copeia, 1951, No. 3, p. 252.

BUDD, J. C.
1952. A Northern Record of the White Crappie from Lake Huron, Ontario. Copeia, 1952, No. 3, p. 210.

CAHN, ALVIN R.
1927. An Ecological Study of Southern Wisconsin Fishes. The Brook Silverside (*Labidesthes sicculus*) and the Cisco (*Leucichthys artedi*) in Their Relations to the Region. Ill. Biol. Monogr., Vol. 11, No. 1, 151 pp., 16 pls.

CARL, G. CLIFFORD, and CLEMENS, W. A.
1953. The Fresh-water Fishes of British Columbia. Second Ed. B. C. Provincial Mus. Handbook, No. 5, 136 pp., 6 figs., 58 text-figs., 7 col. pls.

CARR, ARCHIE, and GOIN, COLEMAN J.
1955. Guide to the Reptiles, Amphibians, and Fresh-water Fishes of Florida. Univ. Florida Press, Gainesville, 341 pp., illus.

CHURCHILL, EDWARD P., and OVER, WILLIAM H.
1933. Fishes of South Dakota. S. D. Dept. Game and Fish, Pierre, 83 pp., 71 figs.

COKER, ROBERT E.
1930. Studies of Common Fishes of the Mississippi River at Keokuk. Bull. U. S. Bur. Fish., Vol. 45, 1929, pp. 141-225, 30 figs.

COOPER, GERALD P.
1939. A Biological Survey of the Waters of York County and the Southern Part of Cumberland County, Maine. Maine Dept. Inland Fish. and Game, Fish Surv. Rept. No. 1, 57 pp., 1 text-fig., 2 maps.

1939. A Biological Survey of Thirty-one Lakes and Ponds of the Upper Saco River and Sebago Lake Drainage Systems in Maine. Maine Dept. Inland Fish. and Game, Fish Surv. Rept. No. 2, 147 pp., 27 figs., 14 pls., 1 map.
1940. A Biological Survey of the Rangeley Lakes, with Special Reference to Trout and Salmon. Maine Dept. Inland Fish. and Game, Fish Surv. Rept. No. 3, 182 pp., 12 figs., 6 pls.
1941. A Biological Survey of Lakes and Ponds of the Androscoggin and Kennebec River Drainage System in Maine. Maine Dept. Inland Fish. and Game, Fish Surv. Rept. No. 4, 238 pp., 56 figs., 25 pls., 1 map, 6 text-figs.
1942. A Biological Survey of Lakes and Ponds of the Central Coastal Area of Maine. Maine Dept. of Inland Fish. and Game, Fish Surv. Rept. No. 5, 184 pp., 59 figs., 9 pls.

Cooper, Gerald P., and Fuller, John L.
1945. A Biological Survey of Moosehead Lake and Haymock Lake, Maine. Maine Dept. Inland Fish. and Game, Fish Surv. Rept. No. 6, 160 pp., 9 figs., 26 pls., 1 map.

Cope, E. D.
1864. Partial Catalogue of the Cold-blooded Vertebrata of Michigan. Pt. I. Proc. Acad. Nat. Sci. Phila., 1864, pp. 276-285.

Cox, Philip
1899. Fresh Water Fishes and Batrachia of Gaspé, P. Q., and their Distribution in the Maritime Provinces of Canada. Trans. Roy. Soc. Canada, Vol. 5, Sect. 4, pp. 141-154.

Creaser, Charles W.
1926. The Structure and Growth of the Scales of Fishes in Relation to the Interpretation of Their Life-history, with Special Reference to the Sunfish, *Eupomotis gibbosus*. Misc. Publ. Mus. Zool. Univ. Mich., No. 17, 82 pp., 12 figs., 1 pl.

Creaser, Charles W., and Hann, Clare S.
1929. The Food of Larval Lampreys. Papers Mich. Acad. Sci., Arts and Letters, Vol. 10, 1928, pp. 433-437.

Creaser, Charles W., and Hubbs, Carl L.
1922. A Revision of the Holarctic Lampreys. Occ. Papers Mus. Zool. Univ. Mich., No. 120, 14 pp., 1 pl.

Cross, Frank B.
1953. Nomenclature in the Pimephalinae, with Special Reference to the Bullhead Minnow, *Pimephales vigilax perspicuus* (Girard). Trans. Kans. Acad. Sci., Vol. 56, No. 1, pp. 92-96.

Cross, Frank Bernard, and Moore, George A.
1952. The Fishes of the Poteau River, Oklahoma and Arkansas. Amer. Midland Nat., Vol. 47, No. 2, pp. 396-412.

Cuerrier, J. P., Fry, F. E. J., and Préfontaine, G.
1946. Liste Préliminaire des Poissons de la Région de Montreal et du Lac Saint-Pierre. Naturaliste Canadien, Vol. 73, pp. 17-32, 1 map.

Curtis, Brian
1938. The Life Story of the Fish. D. Appleton-Century Co., New York, xiv + 260 pp., 36 figs., diagrams A-N.

Dence, W. A.
1952. Establishment of White Perch, *Morone americana,* in Central New York. Copeia, 1952, No. 3, pp. 200-201.

Doan, K. H.
1947. Report for 1946 of the Central Fisheries Research Station, Winnipeg, Man. App. 5. In: Ann. Rept. Fish. Res. Bd. Canada, 1946, pp. 40-41.

DUNBAR, M. J., and HILDEBRAND, H. H.
1952. Contribution to the Study of the Fishes of Ungava Bay. Jour. Fish. Res. Bd. Canada, Vol. 9, No. 2, pp. 83-128, 1 fig.

DYMOND, JOHN R.
1922. A Provisional List of the Fishes of Lake Erie. Univ. Toronto Studies, Biol. Ser., No. 20 (Publ. Ont. Fish. Res. Lab., No. 4), pp. 55-73.
1926. The Fishes of Lake Nipigon. Univ. Toronto Studies, Biol. Ser., No. 27 (Publ. Ont. Fish. Res. Lab., No. 27), 108 pp., 11 pls., 1 map.
1928. The Game Fishes of Canada. Canadian Pacific Railway Co., Canada, 45 pp., 15 col. pls. [The author's name is not indicated on the publication.]
1939. The Fishes of the Ottawa Region. Contrib. Roy. Ont. Mus. Zool., No. 15, 43 pp.
1943. The Coregonine Fishes of Northwestern Canada. Contrib. Roy. Ont. Mus. Zool., No. 24, pp. 171-232, 1 fig., 2 maps.
1947. A List of the Freshwater Fishes of Canada East of the Rocky Mountains with Keys. Roy. Ont. Mus. Zool., Misc. Publ. No. 1, 36 pp.
1955. The Introduction of Foreign Fishes in Canada. Proc. Int. Assoc. Theoret. Applied Limnology, Vol. 12, pp. 543-553.

DYMOND, JOHN R., and HART, JOHN L.
1927. The Fishes of Lake Abitibi (Ontario) and Adjacent Waters. Univ. Toronto Studies, Biol. Ser., No. 29 (Publ. Ont. Fish. Res. Lab., No. 28), 19 pp.

DYMOND, J. R., HART, J. L., and PRITCHARD, A. L.
1929. The Fishes of the Canadian Waters of Lake Ontario. Univ. Toronto Studies, Biol. Ser., No. 33 (Publ. Ont. Fish. Res. Lab., No. 37), 35 pp.

DYMOND, J. R., and VLADYKOV, V. D.
1934. The Distribution and Relationship of the Salmonoid Fishes of North America and North Asia. Proc. Fifth Pacific Sci. Cong., pp. 3741-3750, 10 figs.

EDDY, SAMUEL, and SURBER, THADDEUS
1947. Northern Fishes with Special Reference to the Upper Mississippi Valley. Univ. Minnesota Press, Rev. Ed., 276 pp., 57 figs., some in col.

ELSER, HAROLD J.
1950. The Common Fishes of Maryland; How to Tell Them Apart. Md. Bd. Nat. Res. Dept. Res. and Ed., Publ. No. 88, 45 pp., many text-figs.

ESCHMEYER, PAUL H.
1955. The Reproduction of Lake Trout in Southern Lake Superior. Trans. Amer. Fish. Soc., Vol. 84 (1954), pp. 47-74, 3 figs.

ESCHMEYER, PAUL H., and BAILEY, REEVE M.
1955. The Pygmy Whitefish, *Coregonus coulteri,* in Lake Superior. Trans. Amer. Fish. Soc., Vol. 84 (1954), pp. 161-199, 5 figs.

EVERHART, W. HARRY
1950. Fishes of Maine. Maine Dept. Inland Fish and Game, [ii] + 53 pp., 65 text-figs., 9 col. pls.

EVERMANN, BARTON W.
1902. List of Species of Fishes Known to Occur in the Great Lakes or Their Connecting Waters. Bull. U. S. Fish Comm., Vol. 21, 1901, pp. 95-96.
1918. The Fishes of Kentucky and Tennessee: A Distributional Catalogue of the Known Species. Bull. U. S. Bur. Fish., Vol. 35, 1915-16, pp. 295-368.

EVERMANN, BARTON W., and CLARK, HOWARD W.
1920. Lake Maxinkuckee. A Physical and Biological Survey. Indiana Dept. Cons., Indianapolis, Vol. 1, 660 pp., front., 23 figs., 36 pls. (many col.), 1 map.

EVERMANN, BARTON W., and COX, ULYSSES O.
1896. A Report upon the Fishes of the Missouri River Basin. Rept. U. S. Fish Comm., 1894, pp. 325-429.

EVERMANN, BARTON WARREN, and GOLDSBOROUGH, EDMUND LEE
1907. A Check List of the Freshwater Fishes of Canada. Proc. Biol. Soc. Wash., Vol. 20, pp. 89-119.

EVERMANN, BARTON W., and KENDALL, WILLIAM C.
1894. The Fishes of Texas and the Rio Grande Basin, Considered Chiefly with Reference to Their Geographic Distribution. Bull. U. S. Fish Comm., 1892, pp. 55-126, pls. 10-50.
1900. Check List of the Fishes of Florida. Rept. U. S. Fish Comm., 1899, pp. 37-103.
1902. Notes on the Fishes of Lake Ontario. Rept. U. S. Fish Comm., 1901, pp. 209-216.
1902. An Annotated List of the Fishes Known to Occur in Lake Champlain and its Tributary Waters. Rept. U. S. Fish Comm., 1901, pp. 217-225.
1902. An Annotated List of the Fishes Known to Occur in the St. Lawrence River. Rept. U. S. Fish Comm., 1901, pp. 227-240.

EVERMANN, BARTON WARREN, and LATIMER, HOMER BARKER
1910. The Fishes of the Lake of the Woods and Connecting Waters. Proc. U. S. Nat. Mus., Vol. 39, pp. 121-136.

FISH, MARIE P.
1927. Contribution to the Embryology of the American Eel (*Anguilla rostrata* LeSueur). Zoologica, Vol. 8, No. 5, pp. 289-324, figs. 103-116.
1932. Contributions to the Early Life Histories of Sixty-two Species of Fishes from Lake Erie and Its Tributary Waters. Bull. U. S. Bur. Fish., Vol. 47, 1931-33, pp. 293-398, 44 figs.

FORBES, STEPHEN A., and RICHARDSON, ROBERT E.
1920. The Fishes of Illinois. (Second edit.) Nat. Hist. Surv. Ill., Vol. 3, Text, cxxi + 357 pp., 76 figs., many col. pls., Atlas, 103 maps.

FOWLER, HENRY W.
1906. The Fishes of New Jersey. Ann. Rept. New Jersey State Mus., 1905, Pt. 2, pp. 35-477, 81 figs., front., 103 pls.
1907. A Supplementary Account of the Fishes of New Jersey. Ann. Rept. New Jersey State Mus., 1906, Pt. 3, pp. 251-384, 44 figs., pls. 70-122.
1918. A Review of the Fishes Described in Cope's Partial Catalogue of the Cold-blooded Vertebrata of Michigan. Occ. Papers Mus. Zool. Univ. Mich., No. 60, 51 pp., 13 pls.
1919. A List of the Fishes of Pennsylvania. Proc. Biol. Soc. Wash., Vol. 32, pp. 49-73.
1945. A Study of the Fishes of the Southern Piedmont and Coastal Plain. Phila. Acad. Sci., iv + 408 pp., 313 figs.

FREEMAN, H. W.
1952. New Distribution Records for Fishes of the Savannah River Basin, South Carolina. Copeia, 1952, No. 4, p. 269.

FUNK, J. L., LOWRY, E. M., PATRIARCHE, M. H., MARTIN, R. G., CAMPBELL, R. S., and O'CONNELL, T. R., JR.
1953. The Black River Studies. Univ. Mo. Studies, Vol. 26, No. 2, 136 pp., illus.

GAGE, SIMON H.

1893. The Lake and Brook Lampreys of New York. Especially Those of Cayuga and Seneca Lakes. In: Wilder Quarter-Century Book, Ithaca, pp. 421-493, 8 pls.

1928. The Lampreys of New York State—Life History and Economics. In: A Biological Survey of the Oswego River System. Suppl. 17th Ann. Rept. N. Y. St. Cons. Dept., 1927, pp. 158-191, 7 figs.

1929. Lampreys and Their Ways. Sci. Monthly, Vol. 28, pp. 401-416, 7 figs.

GERKING, SHELBY D.

1945. The Distribution of the Fishes of Indiana. Invest. Ind. Lakes and Streams, Vol. 3, No. 1, 137 pp., 113 maps.

1955. Key to the Fishes of Indiana. Invest. Ind. Lakes and Streams, Vol. 4, pp. 49-86.

GILL, THEODORE N.

1904. A Remarkable Genus of Fishes—the Umbras. Smiths. Misc. Coll., Vol. 45, pp. 295-305, figs. 34-38.

1907. Parental Care Among Fresh-water Fishes. Ann. Rept. Smiths. Inst., 1905, pp. 403-531, 98 figs., 1 pl.

1908. The Miller's-thumb and Its Habits. Smiths. Misc. Coll., Vol. 52, pp. 101-116, figs. 26-39.

GOODE, G. BROWN

1888. American Fishes. (Second edit.) Standard Book Co., New York, xv + 496 pp., col. front., 134 figs.

GREELEY, JOHN R.

1927. Fishes of the Genesee Region with Annotated List. In: A Biological Survey of the Genesee River System. Suppl. 16th Ann. Rept. N. Y. St. Cons. Dept., 1926, pp. 47-66, 8 pls.

1928. Fishes of the Oswego Watershed [with Annotated List]. In: A Biological Survey of the Oswego River System. Suppl. 17th Ann. Rept. N. Y. St. Cons. Dept., 1927, pp. 84-107, 12 col. pls.

1929. Fishes of the Erie-Niagara Watershed [with Annotated List]. In: A Biological Survey of the Erie-Niagara System. Suppl. 18th Ann. Rept. N. Y. St. Cons. Dept., 1928, pp. 150-179, 8 col. pls. 4 illust.

1930. Fishes of the Lake Champlain Watershed [with Annotated List]. In: A Biological Survey of the Champlain Watershed. Suppl. 19th Ann. Rept. N. Y. St. Cons. Dept., 1929, pp. 44-87, 16 col. pls.

1934. Fishes of the Raquette Watershed with Annotated List. In: A Biological Survey of the Raquette Watershed. Suppl. 23rd Ann. Rept. N. Y. St. Cons. Dept., 1933, pp. 53-108, 4 figs., 18 illust., 12 col. pls.

1935. Fishes of the Watershed with Annotated List. In: A Biological Survey of the Mohawk-Hudson Watershed. Suppl. 24th Ann. Rept. N. Y. St. Cons. Dept., 1934, pp. 63-101, 4 col. pls., 5 illust.

1936. Fishes of the Area with Annotated List. In: A Biological Survey of the Delaware and Susquehanna Watersheds. Suppl. 25th Ann. Rept. N. Y. St. Cons. Dept., 1935, pp. 45-88, 4 col. pls., 9 illust.

1937. Fishes of the Area with Annotated List. In: A Biological Survey of the Lower Hudson Watershed. Suppl. 26th Ann. Rept. N. Y. St. Cons. Dept., 1936, pp. 45-103, 3 figs., 4 col. pls., 10 illust.

1938. Fishes of the Area with Annotated List. In: A Biological Survey of the Allegheny and Chemung Watersheds. Suppl. 27th Ann. Rept. N. Y. St. Cons. Dept., 1937, pp. 48-73, 2 col. pls., 3 illust.

1939. The Freshwater Fishes of Long Island and Staten Island with Annotated List. In: A Biological Survey of the Fresh Waters of Long Island. Suppl. 28th Ann. Rept. N. Y. St. Cons. Dept., 1938, pp. 29-44.

1940. Fishes of the Watershed with Annotated List. In: A Biological Survey of the Lake Ontario Watershed. Suppl. 29th Ann. Rept. N. Y. St. Cons. Dept., 1939, pp. 42-81, 4 col. pls., 2 illust.

Greeley, John R., and Bishop, Sherman C.
1932. Fishes of the Area with Annotated List. In: A Biological Survey of the Oswegatchie and Black River Systems. Suppl. 21st Ann. Rept. N. Y. St. Cons. Dept., 1931, pp. 54-92, 3 figs., 12 col. pls., sev. illust.
1933. Fishes of the Upper Hudson Watershed with Annotated List. In: A Biological Survey of the Upper Hudson Watershed. Suppl. 22nd Ann. Rept. N. Y. St. Cons. Dept., 1932, pp. 64-101, 9 figs., 12 col. pls.

Greeley, John R., and Greene, C. Willard
1931. Fishes of the Area with Annotated List. In: A Biological Survey of the St. Lawrence Watershed (Including the Grass, St. Regis, Salmon, Chateaugay Systems and the St. Lawrence between Ogdensburg and the International Boundary). Suppl. 20th Ann. Rept. N. Y. St. Cons. Dept., 1930, pp. 44-94, 12 col. pls.

Greene, C. Willard
1935. The Distribution of Wisconsin Fishes. Wisc. Cons. Comm., Madison, 235 pp., 96 maps.

Gunter, Gordon, and Knapp, Frank T.
1951. Fishes, New, Rare or Seldom Recorded from the Texas Coast. Tex. Jour. Sci., Vol. 3, No. 1, pp. 134-138.

Halkett, Andrew
1913. Check List of the Fishes of the Dominion of Canada and Newfoundland. C. H. Parmalee, Ottawa, 138 pp., 14 pls.

Hall, Gordon E.
1956. Additions to the Fish Fauna of Oklahoma with a Summary of Introduced Species. Southwestern Nat., Vol. 1, No. 1, pp. 16-26.

Halstead, B. W.
1953. The Orthopedic Arthrometer for the Precise Measurement of Angles in Systematic Ichthyology. Copeia, 1953, No. 4, p. 241.

Hankinson, Thomas L.
1924. The Habitat of the Brook Trout in Michigan. Papers Mich. Acad. Sci., Arts and Letters, Vol. 2, 1922, pp. 197-205.
1929. Fishes of North Dakota. Papers Mich. Acad. Sci., Arts and Letters, Vol. 10, 1928, pp. 439-460, pls. 26-29.
1932. Observations on the Breeding Behavior and Habitats of Fishes in Southern Michigan. Papers Mich. Acad. Sci., Arts and Letters, Vol. 15, 1931, pp. 411-425, pls. 33-34.

Harkness, William J. K., and Hart, John L.
1927. The Fishes of Long Lake, Ontario. Univ. Toronto Studies, Biol. Ser., No. 29 (Publ. Ont. Fish. Res. Lab., No. 29), pp. 21-31.

Harmic, Jay L.
1952. Fresh Water Fisheries Survey. Dela. Bd. Game and Fish Comm., Fish. Publ. No. 1, [iv] + 154 pp., many text-figs, end-paper map.

Hay, Oliver P.
1894. The Lampreys and Fishes of Indiana. In: 19th Ann. Rept. Indiana Dept. Geol. and Nat. Res., pp. 147-296.

Henshall, James A.
1906. A List of the Fishes of Montana with Notes on the Game Fishes. Bull. Univ. Montana, No. 34, 10 pp.
1917. Book of the Black Bass. Stewart and Kidd Co., Cincinnati, 452 pp.
1919. Bass, Pike and Perch and Other Game Fishes of America. Stewart and Kidd Co., Cincinnati, 410 pp.

Hile, Ralph
1952. 25 Years of Federal Fishery Research on the Great Lakes. U. S. Fish and Wildlife Service, Special Scientific Report: Fisheries No. 85, 48 pp.

HINKS, DAVID
1943. The Fishes of Manitoba. Manitoba Department of Mines and Natural Resources, Winnipeg, x + 102 pp., 93 figs., 5 maps.

HUBBS, CARL L.
1920. A Comparative Study of the Bones Forming the Opercular Series of Fishes. Jour. Morph., Vol. 33, No. 1, pp. 66-71.
1921. An Ecological Study of the Life-history of the Fresh-water Atherine Fish *Labidesthes sicculus*. Ecology, Vol. 2, No. 4, pp. 262-276, 4 figs.
1925. The Life-cycle and Growth of Lampreys. Papers Mich. Acad. Sci., Arts and Letters, Vol. 4, 1924, pp. 587-603, figs. 16-22.
1926. A Check-list of the Fishes of the Great Lakes and Tributary Waters with Nomenclatorial Notes and Analytical Keys. Misc. Pub. Mus. Zool. Univ. Mich., No. 15, 77 pp., 4 pls.
1929. The Fishes. In: The Book of Huron Mountain. Huron Mountain Club, pp. 153-164, 1 pl.
1930. Further Additions and Corrections to the List of the Fishes of the Great Lakes and Tributary Waters. Papers Mich. Acad. Sci., Arts and Letters, Vol. 11, 1929, pp. 425-436.
1930. Materials for a Revision of the Catostomid Fishes of Eastern North America. Misc. Publ. Mus. Zool. Univ. Mich., No. 20, 47 pp., front.
1931. *Parexoglossum laurae*, a New Cyprinid Fish from the Upper Kanawha River System. Occ. Papers Mus. Zool. Univ. Mich., No. 234, 12 pp., 2 pls.
1945. Corrected Distributional Records for Minnesota Fishes. Copeia, 1945, No. 1, pp. 13-22.

HUBBS, CARL L., and ALLEN, E. ROSS
1944. Fishes of Silver Springs, Florida. Proc. Fla. Acad. Sci., Vol. 6, pp. 110-130, 4 figs.

HUBBS, CARL L., and BAILEY, REEVE M.
1938. The Small-mouthed Bass. Bull. Cranbrook Inst. Sci., No. 10, 89 pp., front., 5 figs., 9 pls.
1940. A Revision of the Black Basses (*Micropterus* and *Huro*) with Descriptions of Four New Forms. Misc. Publ. Mus. Zool. Univ. Mich., No. 48, 51 pp., 1 fig., 6 pls., 2 maps.

HUBBS, CARL L., and BLACK, JOHN D.
1945. Revision of *Ceratichthys*, a Genus of American Cyprinid Fishes. Misc. Publ. Mus. Zool. Univ. Mich., No. 66, 56 pp., 1 fig., 2 pls., 2 maps.

HUBBS, CARL L., and BROWN, DUGALD E. S.
1929. Materials for a Distributional Study of Ontario Fishes. Trans. Roy. Canad. Inst., Vol. 17, Pt. 1, 56 pp.

HUBBS, CARL L., and CANNON, MOTT DWIGHT
1935. The Darters of the Genera *Hololepis* and *Villora*. Misc. Publ. Mus. Zool. Univ. Mich., No. 30, 93 pp., 3 pls.

HUBBS, CARL L., and COOPER, GERALD P.
1936. Minnows of Michigan. Bull. Cranbrook Insti. Sci., No. 8, 95 pp., 2 figs., 10 pls.

HUBBS, CARL L., and CROWE, WALTER R.
1956. Preliminary Analysis of the American Cyprinid Fishes, Seven New, Referred to the Genus *Hybopsis*, subgenus *Erimystax*. Occ. Pap. Mus. Zool. Univ. Mich., No. 578, 8 pp.

HUBBS, CARL L., and GREENE, C. WILLARD
1928. Further Notes on the Fishes of the Great Lakes and Tributary Waters. Papers Mich. Acad. Sci., Arts and Letters, Vol. 8, 1927, pp. 371-392.

HUBBS, CARL L., and LAGLER, KARL F.
1939. Keys for the Identification of the Fishes of the Great Lakes and Tributary Waters. Ann Arbor, published by the authors, 37 pp., 7 figs.
1941. Guide to the Fishes of the Great Lakes and Tributary Waters. Bull. Cranbrook Inst. Sci., No. 18, 100 pp., 118 figs., 1 map.
1943. Annotated List of the Fishes of Foots Pond, Gibson County, Indiana. Invest. Ind. Lakes and Streams, Vol. 2, 1942, No. 4, pp. 73-83.
1947. Fishes of the Great Lakes Region. Bull. Cranbrook Inst. Sci. No. 26, [xi] + 186 pp., 251 figs., 26 col. pls., 38 text-figs., end-paper map. (Corrected second printing, 1949.)
1949. Fishes of Isle Royale, Lake Superior, Michigan. Papers Mich. Acad. Sci., Arts and Letters, Vol. 33 (1947), pp. 73-133, 2 pls., 1 fig. (map).
1957. List of Fishes of the Great Lakes and Tributary Waters. Dept. Fish. Univ. Mich., Mich. Fisheries, No. 1, 6 pp.

HUBBS, CARL L., and RANEY, EDWARD C.
1944. Systematic Notes on North American Siluroid Fishes of the Genus *Schilbeodes*. Occ. Pap. Mus. Zool. Univ. Mich., No. 487, 36 pp., 1 pl., 1 map.

HUBBS, CARL L., and TRAUTMAN, MILTON B.
1937. A Revision of the Lamprey Genus *Ichthyomyzon*. Misc. Publ. Mus. Zool. Univ. Mich., No. 35, 109 pp., 5 figs., 2 pls., 1 map.

HUBBS, CLARK
1951. Records from East Texas of Three Species of Fish, *Semotilus atromaculatus, Notropis cornuta,* and *Microperca proelearis*. Tex. Jour. Sci., Vol. 3, No. 3, p. 490.
1952. Records from East Texas of Three Species of Fish, *Hadropterus maculatus, Etheostoma histrio,* and *Etheostoma barratti*. Tex. Jour. Sci., Vol. 4, No. 4, p. 486.
1954. Corrected Distributional Records for Texas Fresh-water Fishes. Tex. Jour. Sci., Vol. 6, No. 3, pp. 277-291.

HUBBS, CLARK, and HERZOG, WILLIAM F.
1955. The Distribution of the Suckermouth Minnow, *Phenacobius mirabilis*, in Texas. Tex. Jour. Sci., Vol. 7, No. 1, pp. 69-71.

HUBBS, CLARK, KUEHNE, R. A., and BALL, J. C.
1953. The Fishes of the Upper Guadalupe River, Texas. Tex. Jour. Sci., Vol. 5, No. 2, pp. 216-244, 34 figs.

JORDAN, DAVID STARR
1877. Partial Synopsis of the Fishes of Upper Georgia; with Supplementary Papers on Fishes of Tennessee, Kentucky, and Indiana. Ann. New York Lyceum Nat. Hist., Vol. 11, 1874-1877, pp. 307-377.
1903. How to Collect Fishes. The Popular Science Monthly, Vol. 62, No. 1, pp. 85-88.
1905. A Guide to the Study of Fishes. Henry Holt and Company, New York, Vol. 1, xxvi + 624 pp., front., 393 figs; Vol. 2, xxii + 599 pp., front., 506 figs.
1923. A Classification of Fishes. Stanford Univ. Publ., Biol. Sci., Vol. 3, No. 2, pp. 77-243 + x.
1925. Fishes. (Revised edit.) D. Appleton and Co., New York, xvi + 773 pp., col. front., 673 figs., 17 col. pls.
1929. Manual of the Vertebrate Animals of the Northeastern United States, Inclusive of Marine Species. (13th edit.) World Book Co., Yonkers-on-Hudson, New York, xxxi + 446 pp.

JORDAN, DAVID STARR, and BRAYTON, ALEMBERT W.
1878. Contributions to North American Ichthyology No. 3. A—On the Distribution of the Fishes of the Allegheny Region of South Carolina, Georgia, and Tennessee, with Descriptions of New or Little Known Species. Bull. U. S. Nat. Mus., No. 12, 95 pp.

JORDAN, DAVID STARR, and EVERMANN, BARTON W.
1896-1900. The Fishes of North and Middle America. Bull. U. S. Nat. Mus., No. 47, In 4 Parts, 3313 pp., 392 pls.
1902. American Food and Game Fishes. A Popular Account of All the Species Found in America North of the Equator, with Keys for Ready Identification, Life Histories and Methods of Capture. Doubleday, Page and Co., New York, 1 + 573 pp., col. front., 221 figs., 66 pls., 9 col. pls.

JORDAN, DAVID STARR, EVERMANN, BARTON WARREN, and CLARK, HOWARD WALTON
1930. Check List of the Fishes and Fishlike Vertebrates of North and Middle America North of the Northern Boundary of Venezuela and Colombia. Rept. U. S. Comm. Fish., 1928, Pt. 2, iv + 670 pp.

KELEHER, J. J.
1952. Notes on Fishes Collected from Lake Winnipeg Region. Canad. Field-Nat., Vol. 66, No. 6, pp. 170-173.
1956. The Northern Limits of Distribution in Manitoba for Cyprinid Fishes. Canad. Jour. Zool., Vol. 34, pp. 263-266.

KENDALL, WILLIAM C.
1908. Fauna of New England. 8. List of the Pisces. Occ. Papers Bost. Soc. Nat. Hist., Vol. 7, [viii] + 152 pp.
1909. The Fishes of Labrador. Proc. Portland Soc. Nat. Hist., Vol. 2, pp. 207-243.
1910. American Catfishes: Habits, Culture, and Commercial Importance. Rept. U. S. Comm. Fish., 1908, 39 pp., 10 pls.
1914. The Fishes of New England: The Salmon Family. Part 1. The Trout or Charrs. Mem. Bost. Soc. Nat. Hist., Vol. 8, No. 1, 103 pp., 7 col. pls.
1935. The Fishes of New England. The Salmon Family. Part 2. The Salmons. Mem. Bost. Soc. Nat. Hist., Vol. 9, Pt. 1, 166 pp., 11 col. pls.

KENDALL, WILLIAM C., and DENCE, WILFRED A.
1929. The Fishes of the Cranberry Lake Region. Roosevelt Wild Life Bull., Vol. 5, No. 2, pp. 219-309, figs. 63-92.

KENNEDY, W. A.
1953. The Morphometry of the Coregonine Fishes of Great Bear Lake, N. W. T. Jour. Fish Res. Bd. Canad., Vol. 10, No. 2, pp. 51-61.

KNAPP, FRANK T.
1951. Additional Reports of Lampreys from Texas. Copeia, 1951, No. 1, p. 87.
1953. Fishes Found in the Freshwaters of Texas. Ragland Studio and Litho Printing Co., Brunswick, Ga., viii + 166 pp., 190 figs., 1 map.

KOELZ, WALTER
1929. Coregonid Fishes of the Great Lakes. Bull. U. S. Bur. Fish., Vol. 43, 1927, Pt. 2, pp. 297-643, 31 figs.
1931. The Coregonid Fishes of Northeastern North America. Papers Mich. Acad. Sci., Arts and Letters, Vol. 13, 1930, pp. 303-432, 1 pl.

KYLE, HARRY M.
1926. The Biology of Fishes. The Macmillan Co., New York, xvi + 396 pp., 77 figs. 17 pls.

LAMONTE, FRANCESCA
1945. North American Game Fishes. Doubleday, Doran, Garden City, xiv + 202 pp., 71 pls., many in col.

LARSEN, ALFRED
1954. First Record of the White Perch (*Morone americana*) in Lake Erie. Copeia, 1954, No. 2, p. 154.

LEGENDRE, VIANNEY
1942. Redécouverte après un Siécle et Reclassification d'une Espèce de Catostomidé. Canad. Nat., Vol. 69, pp. 227-233, 2 figs.
1951. List of the Freshwater Fishes of the Province of Quebec. Quebec Biol. Bur., 4 pp.
1953. The Freshwater Fishes of the Province of Quebec. In: Ninth Report of the Biological Bureau, Quebec Game and Fisheries Dept., Ottawa, pp. 190-301.
1954. The Freshwater Fishes, Volume 1. Key to Game and Commercial Fishes of the Province of Quebec. Canad. Soc. Ecology and Quebec Game and Fisheries Dept., 180 pp., 80 figs., many text-figs.

LINDSEY, C. C.
1956. Distribution and Taxonomy of Fishes in the Mackenzie Drainage of British Columbia. Jour. Fish. Res. Bd. Canada, Vol. 13, No. 6, pp. 759-789, 2 figs.

LIVINGSTONE, D. A.
1954. The Fresh Water Fishes of Nova Scotia. Proc. N. S. Inst. Sci., Vol. 23, Pt. 1, (1950-51) 90 pp., 36 figs., 43 text-figs.

MARKUS, HENRY C.
1934. Life History of the Blackhead Minnow (*Pimephales promelas*). Copeia, 1934, No. 3, pp. 116-122.

MCCABE, BRITTON C.
1945. Section 3. Fishes. In: Fisheries Survey Report 1942. Mass. Dept. Cons., Boston, pp. 30-68, 5 figs., 7 text-figs.

MCCABE, BRITTON C., and SWARTZ, ALBERT H.
1951. Fisheries Report. Mass. Div. Fish and Game Bur. Wildl. Res. and Mgt., 269 pp., 36 figs., many maps.

MICHAEL, ELLIS L.
1906. Catalogue of Michigan Fish. Bull. Fish Comm., No. 8, 45 pp.

MILLER, ROBERT RUSH
1950. A Review of the American Clupeid Fishes of the Genus *Dorosoma*. Proc. U. S. Nat. Mus., Vol. 100, No. 3267, pp. 387-410.
1952. Bait Fishes of The Lower Colorado River from Lake Mead, Nevada, to Yuma, Arizona, with a Key for Their Identification. Calif. Fish and Game, Vol. 38, No. 1, pp. 7-42, 32 figs.
1955. An Annotated List of the American Cyprinodontid Fishes of the Genus *Fundulus*, with the description of *Fundulus persimilis* from Yucatan. Occ. Papers Mus. Zool. Univ. Mich., No. 568, 27 pp., 2 figs.
1957. Origin and Dispersal of the Alewife, *Alosa pseudoharengus*, and the Gizzard Shad, *Dorosoma cepedianum*, in the Great Lakes. Trans. Amer. Fish. Soc., Vol. 86 (1956), pp. 97-111.

MILLER, ROBERT RUSH, and WINN, HOWARD ELLIOTT
1951. Additions to the Known Fish Fauna of Mexico: Three Species and one Subspecies from Sonora. Jour. Wash. Acad. Sci., Vol. 41, No. 2, pp. 83-84.

MOORE, GEORGE A.
1952. Fishes of Oklahoma. Okla. Game and Fish. Dept., 12 pp., 15 text-figs.

MOORE, GEORGE A., and PADEN, JOHN M.
1950. The Fishes of the Illinois River in Oklahoma and Arkansas. Amer. Midland Nat., Vol. 44, No. 1, pp. 76-95, 1 fig.

MORTON, WM. MARKHAM, and MILLER, ROBERT RUSH
1954. Systematic Position of the Lake Trout, *Salvelinus namaycush*. Copeia, 1954, No. 2, pp. 116-124, 2 figs., 2 pls.

NASH, C. W.
1908. Check List of the Fishes of Ontario. In: Vertebrates of Ontario. Ontario Dept. of Education, Toronto, 122 pp., 8 figs., 32 pls.

NORMAN, JOHN R.
1936. A History of Fishes. (Second edit.) Ernest Benn Limited, London, xv + 463 pp., front., 147 figs., 7 pls.

OKKELBERG, PETER
1922. Notes on the Life-history of the Brook Lamprey, *Ichthyomyzon unicolor.* Occ. Papers. Mus. Zool. Univ. Mich., No. 125, 14 pp., 4 figs.

OSBURN, RAYMOND C., WICKLIFF, EDWARD L., and TRAUTMAN, MILTON B.
1930. A Revised List of the Fishes of Ohio. Ohio Jour. Sci., Vol. 30, pp. 169-176.

RADFORTH, ISOBEL
1944. Some Considerations on the Distribution of Fishes in Ontario. Cont. Roy. Ont. Mus. Zool., No. 25, 116 pp., 32 figs.

RANEY, EDWARD C.
1939. The Distribution of the Fishes of the Ohio Drainage Basin of Western Pennsylvania. Cornell Univ., Abstracts of Theses, 1938, pp. 273-277.
1950. Freshwater Fishes. In: The James River Basin, Past, Present and Future. Va. Acad. Sci., Richmond, Va., pp. 151-194.

RANEY, EDWARD C., and MASSMANN, WILLIAM H.
1953. The Fishes of the Tidewater Section of the Pamunkey River, Virginia. Jour. Wash. Acad. Sci., Vol. 43, No. 12, pp. 424-432.

[RAWSON, D. S.]
1947. The Fishes of Saskatchewan. In: Report of the Royal Commission on the Fisheries of the Province of Saskatchewan. Regina, pp. 15-19.

RAWSON, D. S., and ATTON, F. M.
1953. Biological Investigation and Fisheries Management at Lac la Ronge, Saskatchewan. Sask. Dept. Nat. Res., Fish. Bur., 39 pp., front., 16 figs.

REEVES, CORA D.
1907. The Breeding Habits of the Rainbow Darter (*Etheostoma coeruleum* Storer), a Study in Sexual Selection. Biol. Bull., Vol. 14, pp. 35-59, 3 figs.

REIGHARD, JACOB
1903. The Natural History of *Amia calva* Linnaeus. In: Mark Anniversary Volume, Henry Holt and Co., New York, pp. 57-109, pl. 7.
1906. The Breeding Habits, Development and Propagation of the Black Bass (*Micropterus dolomieu* Lacépède and *Micropterus salmoides* Lacépède). Bull. Mich. Fish Comm., No. 7, 73 pp., figs. A-K and 1-29.
1910. Methods of Studying the Habits of Fishes with an Account of the Breeding Habits of the Horned Dace. Bull. U. S. Bur. Fish., Vol. 28, 1908, pp. 1113-1136, 5 figs., pls. 114-120.
1913. The Breeding Habits of the Log-perch (*Percina caprodes*). 15th Ann. Rept. Mich. Acad. Sci., pp. 104-105.
1915. An Ecological Reconnoissance of the Fishes of Douglas Lake, Cheboygan County, Michigan, in Midsummer. Bull. U. S. Bur. Fish., Vol. 33, 1913, pp. 215-249, 4 figs.
1920. The Breeding Behavior of the Suckers and Minnows. Biol. Bull., Vol. 38, 32 pp.

ROBINS, C. RICHARD, and RANEY, EDWARD C.
1956. Studies of the Catostomid Fishes of the Genus *Moxostoma,* with Descriptions of Two New Species. Cornell Univ. Agric. Exp. Sta. Mem. 343, 56 pp., 5 pls.
1957. The Systematic Status of the Suckers of the Genus *Moxostoma* from Texas, New Mexico and Mexico. Tulane Stud. Zool., Vol. 5, No. 12, pp. 291-318.
1957. Distributional and Nomenclatorial Notes on the Suckers of the Genus *Moxostoma.* Copeia, 1957, No. 2, pp. 154-155.

ROSTLUND, ERHARD
1952. Freshwater Fish and Fishing in Native North America. Univ. Calif. Press, Berkeley, x + 313 pp., 47 maps.

SCHRENKEISN, RAY
1938. Field Book of Fresh-water Fishes of North America North of Mexico. G. P. Putnam's Sons, New York, xii + 312 pp.

SCHULTZ, LEONARD P.
1930. The Life History of *Lampetra planeri* Bloch, with a Statistical Analysis of Growth of the Larvae from Western Washington. Occ. Papers Mus. Zool. Univ. Mich., No. 221, 35 pp., 2 pls.

SCOTT, D. M.
1955. Occurrence of the Ninespine Stickleback, *Pungitius pungitius,* in Newfoundland, Canada. Copeia, 1955, No. 1, p. 56.
1955. Additional Records of Two Fishes, *Erimyzon sucetta kennerlyi* and *Hadropterus copelandi,* from southern Ontario, Canada. Copeia, 1955, No. 2, p. 151.

SCOTT, W. B.
1954. Freshwater Fishes of Eastern Canada. Univ. Toronto Press, Toronto, xiv + 128 pp., 109 text-figs.

SIMON, JAMES R.
1946. Wyoming Fishes. Bull. Wyo. Game and Fish Dept., No. 4, 129 pp., 92 figs.

SIMON, JAMES R., and SIMON, FELIX
1939. Check List and Keys of the Fishes of Wyoming. Univ. Wyo. Publ., Vol. 6, No. 4, pp. 47-62.

SMITH, BERTRAM G.
1908. The Spawning Habits of *Chrosomus erythrogaster.* Biol. Bull., Vol. 15, pp. 9-18.
1922. Notes on the Nesting Habits of Cottus. Papers Mich. Acad. Sci., Arts and Letters, Vol. 2, pp. 221-224, pl. 11.

SMITH, HUGH M.
1907. The Fishes of North Carolina. North Carolina Geol. and Econ. Survey, Vol. 2, xi + 453 pp., 188 figs., 21 pls.

SMITH, HUGH M., and HARRON, L. G.
1903. Breeding Habits of the Yellow Catfish. Bull. U. S. Fish. Comm., Vol. 22, 1902, pp. 149-154.

SMITH, O. W.
1922. The Book of the Pike. Stewart Kidd Co., Cincinnati, 197 pp., 13 pls.

SPEIRS, J. MURRAY
1952. Nomenclature of the Channel Catfish and the Burbot in North America. Copeia, 1952, No. 2, pp. 99-103, 1 fig.
1953. History of the Original Descriptions of Great Lakes Fishes. Univ. Toronto, Dept. Zool., Ont. Fish Res. Lab., Multilithed. 38 pp.

STEWART, NORMAN H.
1926. Development, Growth, and Food Habits of the White Sucker, *Catostomus commersonii* LeSueur. Bull. U. S. Bur. Fish., No. 42, pp. 147-184, 55 figs.

SURBER, THADDEUS
1920. A Preliminary Catalogue of the Fishes and Fish-like Vertebrates of Minnesota. Appendix Bienn. Rept. State Game and Fish Comm., 1920, 92 pp., 66 figs.

TAYLOR, WILLIAM RALPH
1954. Records of Fishes in the John N. Lowe Collection from the Upper Peninsula of Michigan. Misc. Publ. Mus. Zool. Univ. Mich. No. 87, 50 pp.

TESTER, ALBERT L.
1931. Spawning Habits of the Small-mouthed Black Bass in Ontario Waters. Trans. Amer. Fish. Soc., Vol. 60, 1930, 9 pp., 3 figs.
1932. Food of the Small-mouthed Black Bass (*Micropterus dolomieu*) in some Ontario Waters. Univ. Toronto Studies, Biol. Ser., No. 36 (Publ. Ont. Fish. Res. Lab., No. 46), pp. 169-203.
1932. Rate of Growth of the Small-mouthed Black Bass (*Micropterus dolomieu*) in some Ontario Waters. Univ. Toronto Studies, Biol. Ser., No. 36 (Publ. Ont. Fish. Res. Lab., No. 47), pp. 205-221.

[TRAUTMAN, MILTON B.]
1940. Artificial Keys for the Identification of the Fishes of the State of Ohio. The Franz Theodore Stone Laboratory, Put-in-Bay, Ohio, iii + 38 pp. Mimeographed. [Rev. edit. mimeo. 1946]

TRAUTMAN, MILTON B.
1956. *Carpiodes cyprinus hinei*, a New Subspecies of Carpsucker from the Ohio and Upper Mississippi River Systems. Ohio Jour. Sci., Vol. 56, No. 1, pp. 33-40, 1 map, 2 text-figs.
1957. The Fishes of Ohio. Ohio State Univ. Press, Columbus, xviii + 683 pp.

TRAUTMAN, MILTON B., and MARTIN, ROBERT G.
1951. *Moxostoma aureolum pisolabrum*, a New Subspecies of Sucker from the Ozarkian Streams of the Mississippi River System. Occ. Pap. Mus. Zool, Univ. Mich. No. 534, 10 pp., 4 figs., 1 pl.

TRAVER, JAY R.
1929. The Habits of the Black-nosed Dace, *Rhinichthys atronasus* (Mitchill). Jour. Elisha Mitchell Sci. Soc., Vol. 45, No. 1, pp. 101-120.

TRUITT, R. V., BEAN, B. A., and FOWLER, H. W.
1929. The Fishes of Maryland. Md. Cons. Dept., Cons. Bull. No. 3, 120 pp., 62 figs., col. front.

VAN CLEAVE, HARLEY J., and MARKUS, HENRY C.
1929. Studies on the Life History of the Blunt-nosed Minnow. Amer. Nat., Vol. 63, pp. 530-539.

VAN OOSTEN, JOHN
1958. Great Lakes Fauna, Flora and their Environment. A Bibliography. Great Lakes Commission, Ann Arbor, x + 86 pp.

VLADYKOV, VADIM D.
1942. Two Fresh-water Fishes New for Quebec. Copeia, 1942, No. 3, pp. 193-194.
1945. Trois Poissons Nouveaux pour la Province de Québec. Nat. Canad., Vol. 72, Nos. 1 and 2, pp. 27-39, 5 figs.

VLADYKOV, VADIM D.
1949. Quebec lampreys (Petromyzonidae) I. List of Species and Their Economical Importance. Quebec Dept. Fish., Contrib. No. 26, 67 pp., front., 21 figs.
1950. Larvae of Eastern American Lampreys (Petromyzonidae) I.—Species with Two Dorsal Fins. Nat. Canad., Vol. 77, Nos. 3-4, pp. 73-95, 13 figs.
1952. Distribution des Lamproies (Petromyzonidae) dans la Province de Québec. Nat. Canad., Vol. 79, pp. 85-120, 13 figs.
1952. Présence dans le Québec du *Morone americana,* Troisième Espèce des Serranidés. Nat. Canad., Vol. 79, No. 12, pp. 325-329, 1 fig.
1954. Taxonomic Characters of the Eastern North American Chars (*Salvelinus* and *Cristivomer*). Jour. Fish. Res. Bd. Canada, Vol. 11, No. 6, pp. 904-932, 12 figs.

WALTERS, VLADIMIR
1955. Fishes of Western Arctic America and Eastern Arctic Siberia. Taxonomy and Zoogeography. Bull. Amer. Mus. Nat. Hist., Vol. 106, Art. 5, pp. 257-368.

WEBSTER, DWIGHT A.
1942. The Life Histories of Some Connecticut Fishes. In: A Fishery Survey of Important Connecticut Lakes. Conn. St. Geol. and Nat. Hist. Surv. Bull., No. 63, pp. 122-227, 60 figs.

WEED, ALFRED C.
1927. Pike, Pickerel and Muskalonge. Field Mus. Nat. Hist., Zool. Leaflet No. 9, 52 pp., 4 figs., 8 pls.

WEISEL, GEORGE F., and DILLON, JOHN B.
1954. Observations on the Pygmy Whitefish, *Prosopium coulteri,* from Bull Lake, Montana. Copeia, 1954, No. 2, pp. 124-127, 1 pl.

WICKLIFF, E. L., and TRAUTMAN, MILTON B.
1932. Some Food and Game Fishes of Ohio. Bull. Ohio Dept. Agric., Bur. Sci. Res., Div. Cons., No. 7, 40 pp., 14 figs., 14 dist. maps.

WILIMOVSKY, NORMAN J.
1954. List of the Fishes of Alaska. Stanford Ichthyol. Bull., Vol. 4, No. 5, pp. 279-294.
1958. Provisional Keys to the Fishes of Alaska. Juneau, Alaska, Fish. Res. Lab., U. S. Fish Wildl. Service, 113 pp., many figs.

WYNNE-EDWARDS, V. C.
1952. Freshwater Vertebrates of the Arctic and Subarctic. Bull. Fish. Res. Bd. Canada, No. 94, 28 pp., 3 figs.

HALFTONE PLATES

The figures composing the following plates are from several sources. Those which are not our own are acknowledged by the abbreviations given below.

F. N. B.—Frank N. Blanchard

B. S. N. H.—Boston Society of Natural History

C. N. H. M.—Chicago Natural History Museum

I. F. R.—Institute for Fisheries Research, Michigan Department of Conservation

I. N. H. S.—Illinois Natural History Survey Division

V. L.—Vianney Legendre

N. Y. C. D.—New York State Conservation Department, Bureau of Biological Survey

U. M. M. Z.—University of Michigan Museum of Zoology

U. S. F. W. S.—U. S. Fish and Wildlife Service

J. V. C.—Jack Van Coevering

LAMPREY FAMILY—Petromyzontidae

Fig. 51.

Silver lamprey; *Ichthyomyzon unicuspis* Hubbs and Trautman
Adult female. About 0.3 × natural size
Ohio, Lucas Co.

Fig. 52.

Northern brook lamprey
Ichthyomyzon fossor Reighard and Cummins
Female above, male below. About 0.4 × natural size
Mich. (F. N. B.)

Fig. 53.

Chestnut lamprey; *Ichthyomyzon castaneus* Girard
Adult male. About 0.3 × natural size
Ill., Randolph Co.

Fig. 54.

Sea lamprey; *Petromyzon marinus* Linnaeus
Adult female. About 0.2 × natural size
Mich., L. Michigan (I. F. R.)

Fig. 55.

American brook lamprey
Entosphenus lamottenii lamottenii (LeSueur)
Adult male (with papilla). About 0.5 × natural size
Mich. (F. N. B.)

PADDLEFISH FAMILY—Polyodontidae

Fig. 56.

Paddlefish; *Polyodon spathula* (Walbaum)
Adult. About 0.1 or less × natural size
Ill. (I. N. H. S.)

STURGEON FAMILY—Acipenseridae

Fig. 57.

Lake sturgeon
Acipenser fulvescens Rafinesque
Juvenile (see also col. pl. 1)
About 0.2 × natural size; Mich., Lake Huron (I. F. R.)

GAR FAMILY—Lepisosteidae

Fig. 58.

Spotted gar; *Lepisosteus productus* (Cope)
Adult. About 0.1 × natural size
Ind., Gibson Co.

Fig. 59.

Northern longnose gar; *Lepisosteus osseus oxyurus* Rafinesque
Half-grown. About 0.2 × natural size
Mich., Livingston Co. (I. F. R.)

BOWFIN FAMILY—Amiidae

Fig. 60.

Bowfin; *Amia calva* Linnaeus
Young (note caudal ocellus, obsolescent in adult female; for male see also col. pl. 2). About 0.5 × natural size
Mich. (F. N. B.)

MOONEYE FAMILY—Hiodontidae

Fig. 61.

Mooneye
Hiodon tergisus LeSueur
Adult (See also col. pl. 3)
About 0.3 × natural size
Ill. (I.N.H.S.)

HERRING FAMILY—Clupeidae

Fig. 62.

Alewife
Pomolobus pseudoharengus (Wilson)
Adult
About 0.5 × natural size
N. Y., L. Ontario

Fig. 63.

American shad
Alosa sapidissima (Wilson)
Adult male
About 0.16 × natural size
N. Y. (N. Y. C. D.)

Fig. 64.

Gizzard shad
Dorosoma cepedianum (LeSueur)
Adult (see also col. pl. 4)
About 0.3 × natural size
Ill. (I. N. H. S.)

SALMON FAMILY—Salmonidae

Fig. 65.

Landlocked Atlantic salmon
Salmo salar sebago Girard
Adult female
About 0.15 × natural size
Maine, Sebago L. tributary (B. S. N. H.)

Fig. 66.

Landlocked Atlantic salmon
Salmo salar sebago Girard
Adult male
About 0.14 × natural size
Maine, Sebago L. tributary (B. S. N. H.)

Fig. 67.
Brown trout
Salmo trutta fairo Linnaeus
Adult male (see also col. pl. p. xiv)
About 0.2 × natural size
N. Y. (N. Y. C. D.)

Fig. 68.
Coast rainbow trout
Salmo gairdnerii irideus Gibbons
Juvenile (note small parr marks)
About 0.8 × natural size
Mich. (F. N. B.)

Fig. 69.
Coast rainbow trout
Salmo gairdnerii irideus Gibbons
Adult female
About 0.13 × natural size
N.Y. (N.Y.C.D.)

Fig. 70.
Brook trout
Salvelinus fontinalis (Mitchill)
Fingerling (note parr marks grading into color of lower side)
About 0.9 × natural size
Mich. (F. N. B.)

Fig. 71.
Brook trout
Salvelinus fontinalis (Mitchill)
Adult male (see also col. pl. 5)
About 0.4 × natural size
N. Y. (N. Y. C. D.)

Fig. 72.
Common lake trout
Salvelinus namaycush namaycush (Walbaum)
Adult female
About 0.15 × natural size
N. Y. (N. Y. C. D.)

WHITEFISH FAMILY—Coregonidae

Fig. 73.
Great Lakes cisco
Coregonus artedii artedii LeSueur
Adult
About 0.3 × natural size
Mich., L. Huron (U. S. F. W. S.)

Fig. 74.
Ives Lake cisco
Coregonus hubbsi (Koelz)
Female
About 0.5 × natural size
Mich., Marquette Co. (U. M. M. Z.)

Fig. 75.
Nipigon tullibee
Coregonus nipigon (Koelz)
Adult male
About 0.3 × natural size
Ont., L. Nipigon (U. S. F. W. S.)

Fig. 76.
Siskiwit Lake cisco
Coregonus bartletti (Koelz)
Adult
About 0.4 × natural size
Mich., Isle Royale (U. M. M. Z.)

Fig. 77.
Michigan shortnose cisco
Coregonus reighardi reighardi (Koelz)
Adult female
About 0.4 × natural size
Ind., L. Michigan (U. S. F. W. S.)

Fig. 78.
Shortjaw cisco
Coregonus zenithicus (Jordan and Evermann)
Adult male
About 0.3 × natural size
Wis., L. Superior (U. S. F. W. S.)

Fig. 79.
Longjaw cisco
Coregonus alpenae (Koelz)
Adult male
About 0.3 × natural size
Mich., L. Michigan (U. S. F. W. S.)

Fig. 80.
Great Lakes bloater
Coregonus hoyi (Gill)
Adult male
About 0.3 × natural size
Wis., L. Michigan (U. S. F. W. S.)

Fig. 81.
Deepwater cisco
Coregonus johannae (Wagner)
Adult male
About 0.3 × natural size
Mich., L. Michigan (U. S. F. W. S.)

Fig. 82.
Michigan kiyi
Coregonus kiyi kiyi (Koelz)
Adult female
About 0.5 × natural size
Wis., L. Michigan (U. S. F. W. S.)

Fig. 83.
Michigan blackfin
Coregonus nigripinnis nigripinnis (Gill)
Adult male
About 0.2 × natural size
Wis., L. Michigan (U. S. F. W. S.)

Fig. 84.
Great Lakes whitefish
Coregonus clupeaformis clupeaformis (Mitchill)
Half-grown
About 0.2 × natural size
Mich., L. Huron (U. S. F. W. S.)

Fig. 85.
Round whitefish
Prosopium cylindraceum quadrilaterale (Richardson)
Adult female
About 0.3 × natural size
Ont., L. Huron (U. S. F. W. S.)

Fig. 86.
Pygmy whitefish
Prosopium coulteri (Eigenmann and Eigenmann)
Adult
About 0.5 × natural size
Mich., L. Superior (U. S. F. W. S.)

GRAYLING FAMILY—Thymallidae

Fig. 87.
Sailfin Arctic grayling
Thymallus arcticus signifer (Richardson)
Adult
About 0.4 × natural size
Mont., Madison Co.

SMELT FAMILY—Osmeridae

Fig. 88.
American smelt
Osmerus mordax (Mitchill)
Adult
About 0.4 × natural size
Mich., Benzie Co. (I. F. R.)

SUCKER FAMILY—Catostomidae

Fig. 89.
Bigmouth buffalo
Ictiobus cyprinellus (Valenciennes)
Adult (see also col. pl. 6)
About 0.15 × natural size
Ill. (I. N. H. S.)

Fig. 90.
Black buffalo
Ictiobus niger (Rafinesque)
Adult
About 0.2 × natural size
Ill. (I. N. H. S.)

Fig. 91.
Smallmouth buffalo
Ictiobus bubalus (Rafinesque)
Immature
About 0.4 × natural size
Mo., Pemiscot Co.

Fig. 92.
Central quillback carpsucker
Carpiodes cyprinus hinei Trautman
Adult
About 0.2 × natural size
Ill. (I. N. H. S.)

Fig. 93.
Northern River carpsucker
Carpiodes carpio carpio (Rafinesque)
Immature male
About 0.4 × natural size
Mo., Caldwell Co.

Fig. 94.
Common white sucker
Catostomus commersonnii commersonnii (Lacépède)
Breeding male (note nuptial tubercles on anal and caudal)
(For non-breeding fish see col. pl. 7)
About 0.3 × natural size
Mich. (F. N. B.)

Fig. 95.
Eastern longnose sucker
Catostomus catostomus catostomus (Forster)
Adult female
About 0.7 × natural size
Mich., Keweenaw Co.

Fig. 96.
Northern hog sucker
Hypentelium nigricans (LeSueur)
Half-grown (for adult see col. pl. 11)
About 0.5 × natural size
Mich. (F. N. B.)

Fig. 97.
Western lake chubsucker
Erimyzon sucetta kennerlii (Girard)
Young (stripe unusually prominent)
About 1.6 × natural size
Mich. (I. F. R.)

Fig. 98.
Western lake chubsucker
Erimyzon sucetta kennerlii (Girard)
Breeding male (note nuptial tubercles and split anal)
About 0.6 × natural size
Mich. (F. N. B.)

Fig. 99.
Western creek chubsucker
Erimyzon oblongus claviformis (Girard)
Adult male (for adult female see col. pl. 9)
About 0.8 × natural size
Ill., Jackson Co.

Fig. 100
Spotted sucker
Minytrema melanops (Rafinesque)
Adult (see also col. pl. 8)
About 0.3 × natural size
Mich., St. Clair (I. F. R.)

Fig. 101.
Black redhorse
Moxostoma duquesnii (LeSueur)
Immature male
About 0.3 × natural size
Mo.

Fig. 102.
Greater redhorse
Moxostoma valenciennesi Jordan
Immature male
About 0.3 × natural size
N. Y. (N. Y. C. D.)

Fig. 103.
Golden redhorse
Moxostoma erythrurum (Rafinesque)
Juvenile
About 0.5 × natural size
Mich., Saginaw Co.

Fig. 104.
Golden redhorse
Moxostoma erythrurum (Rafinesque)
Breeding male (note nuptial tubercles on head)
About 0.2 × natural size
Mich. (F. N. B.)

Fig. 105.
Silver redhorse
Moxostoma anisurum (Rafinesque)
Juvenile
About 0.4 × natural size
Iowa

Fig. 106.
Silver redhorse
Moxostoma anisurum (Rafinesque)
Breeding male (tubercles on anal fin)
About 0.15 × natural size
N. Y. (N. Y. C. D.)

Fig. 107.
Northern shorthead redhorse
Moxostoma macrolepidotum macrolepidotum (LeSueur)
Adult (see also col. pl. 10)
About 0.3 × natural size
Ill. (I. N. H. S.)

Fig. 108.
River redhorse
Moxostoma carinatum (Cope)
Adult
About 0.2 × natural size
Ill. (I. N. H. S.)

Fig. 109.
Copper redhorse
Moxostoma hubbsi Legendre
Adult male
About 0.12 × natural size
Quebec (V. L.)

MINNOW FAMILY—Cyprinidae

Fig. 110.
Carp
Cyprinus carpio Linnaeus
Adult (see also col. pl. 12)
About 0.2 × natural size
Ill. (I. N. H. S.)

Fig. 111.
Goldfish
Carassius auratus (Linnaeus)
Olive-colored wild type
About 0.8 × natural size
Mich. (F. N. B.)

Fig. 112.
Goldfish
Carassius auratus (Linnaeus)
Gold-colored cultivated type with black spots
About 0.3 × natural size
Mich., L. St. Clair (I. F. R.)

Fig. 113.

Fallfish
Semotilus corporalis (Mitchill)
Adult male
About 0.6 × natural size
N. H., Hillsboro Co.

Fig. 114.

Northern creek chub
Semotilus atromaculatus atromaculatus (Mitchill)
Half-grown
About natural size
Mich. (I. F. R.)

Fig. 115.

Northern creek chub
Semotilus atromaculatus atromaculatus (Mitchill)
Mature female (dark specks due to parasites)
About 0.6 × natural size
Mich. (F. N. B.)

Fig. 116.

Northern creek chub
Semotilus atromaculatus atromaculatus (Mitchill)
Breeding male
About 0.3 × natural size
Mich. (I. F. R.)

Fig. 117.

Northern pearl dace
Semotilus margarita nachtriebi (Cox)
Adult male
About 1.1 × natural size
Mich., Schoolcraft Co.

Fig. 118.

Lake northern chub
Hybopsis plumbea plumbea (Agassiz)
Adult female
About 1.1 × natural size
Mich.

Fig. 119.

Creek northern chub
Hybopsis plumbea, subspecies
Adult female
About 0.8 × natural size
Mich., Keweenaw Co.

Fig. 120.
Hornyhead chub
Hybopsis biguttata (Kirtland)
Adult female (note round, blackish caudal spot)
About 0.8 × natural size
Mich. (F. N. B.)

Fig. 121.
Hornyhead chub
Hybopsis biguttata (Kirtland)
Breeding male (tubercles developed to occiput)
About 0.5 × natural size
Mich. (F. N. B.)

Fig. 122.
River chub
Hybopsis micropogon (Cope)
Breeding male (note swollen occiput and tubercles not extending to occiput)
About 0.5 × natural size
Mich. (F. N. B.)

Fig. 123.
Silver chub
Hybopsis storeriana (Kirtland)
Adult male
About 0.8 × natural size
Iowa, Wapello Co.

Fig. 124.
Bigeye chub
Hybopsis amblops (Rafinesque)
Adult female
About 1.5 × natural size
Mich., Monroe Co.

Fig. 125.
Western gravel chub
Hybopsis x-punctata x-punctata Hubbs and Crowe
Adult
About 1.3 × natural size
Mo., Osage Co.
(The eastern subspecies, *trautmani,* has a longer and more decurved snout and a slenderer caudal peduncle)

Fig. 126.
Western blacknose dace
Rhinichthys atratulus meleagris Agassiz
Breeding male (tubercles very small)
About 1.2 × natural size
Mich. (I. F. R.)

Fig. 127.
Great Lakes longnose dace
Rhinichthys cataractae cataractae (Valenciennes)
Adult (note projecting snout)
About natural size
Mich. (F. N. B.)

Fig. 128.
Eastern tonguetied minnow
Parexoglossum laurae laurae Hubbs
Adult male
About 1.2 × natural size
Va., Bland Co.

Fig. 129.
Cutlips
Exoglossum maxillingua (LeSueur)
Adult female
About natural size
N. Y., Monroe Co.

Fig. 130.
Finescale dace
Chrosomus neogaeus (Cope)
Adult male
About 1.3 × natural size
Mich., Keweenaw Co.

Fig. 131.
Northern redbelly dace
Chrosomus eos Cope
Adult (note more oblique mouth and shorter snout)
About 1.4 × natural size
Mich. (I. F. R.)

Fig. 132.
Southern redbelly dace
Chrosomus erythrogaster Rafinesque
Adult (note less oblique mouth and longer snout)
About 1.3 × natural size
Mich. (F. N. B.)

Fig. 133.
Redside dace
Clinostomus elongatus (Kirtland)
Adult female
About 0.9 × natural size
Mich., Wayne Co.

Fig. 134.
Pugnose minnow
Opsopoeodus emiliae Hay
Adult male
About 1.5 × natural size
Mich., Wayne Co.

Fig. 135.
Western golden shiner
Notemigonus crysoleucas auratus (Rafinesque)
Adult (see also col. pl. 13)
About 0.7 × natural size
Mich. (F. N. B.)

Fig. 136.
Lake emerald shiner
Notropis atherinoides acutus (Lapham)
Adult
About natural size
Mich. (I. F. R.)
(River emerald shiner shown on col. pl. 14)

Fig. 137.
Silver shiner
Notropis photogenis (Cope)
Adult female
About 0.7 × natural size
Mich., Washtenaw Co.

Fig. 138.
Rosyface shiner
Notropis rubellus (Agassiz)
Adult
About 1.1 × natural size
Mich. (I. F .R.)

Fig. 139.
Northern redfin shiner
Notropis umbratilis cyanocephalus (Copeland)
Adult
About 1.3 × natural size
Mich., Washtenaw Co.

Fig. 140.
Central common shiner
Notropis cornutus chrysocephalus (Rafinesque)
Adult male (see also col. pl. 15)
About 0.7 × natural size
Mich. (I. F. R.)

Fig. 141.
Northern common shiner
Notropis cornutus frontalis (Agassiz)
Adult female
About natural size
Mich. (F. N. B.)

Fig. 142.
Northern common shiner
Notropis cornutus frontalis (Agassiz)
Breeding male (note color patern)
About 0.7 × natural size
Mich. (F. N. B.)

Fig. 143.
Northern river shiner
Notropis blennius jejunus (Forbes)
Adult female
About 1.4 × natural size
Wis., Vernon Co.

Fig. 144.
Great Lakes spottail shiner
Notropis hudsonius, subspecies
Adult
About 1.2 × natural size
Mich. (F. N. B.)

Fig. 145.
Ironcolor shiner
Notropis chalybaeus (Cope)
Adult male
About 1.8 × natural size
Ind., Newton Co.

Fig. 146.
Northern weed shiner
Notropis roseus richardsoni Hubbs and Greene
Adult male
About 1.7 × natural size
Mich., Allegan Co.

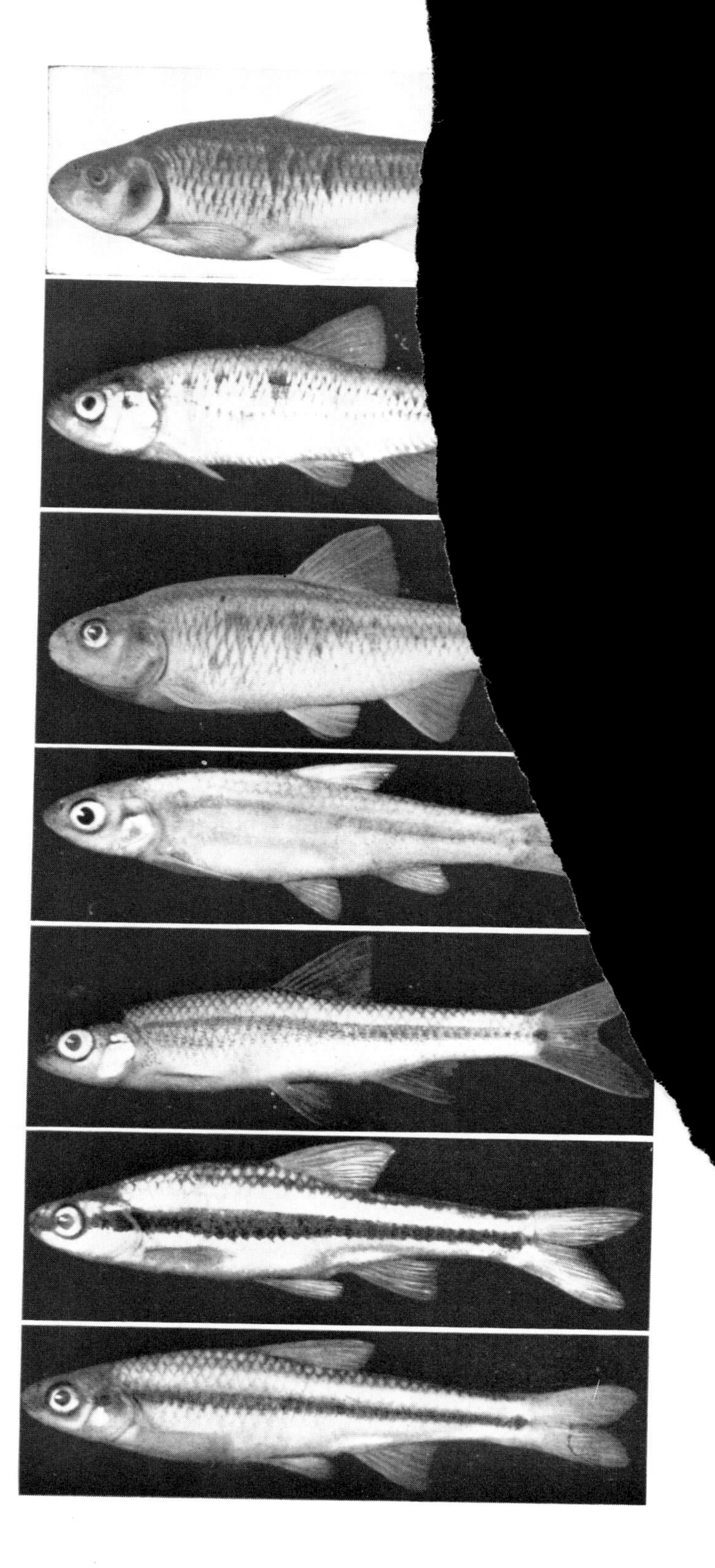

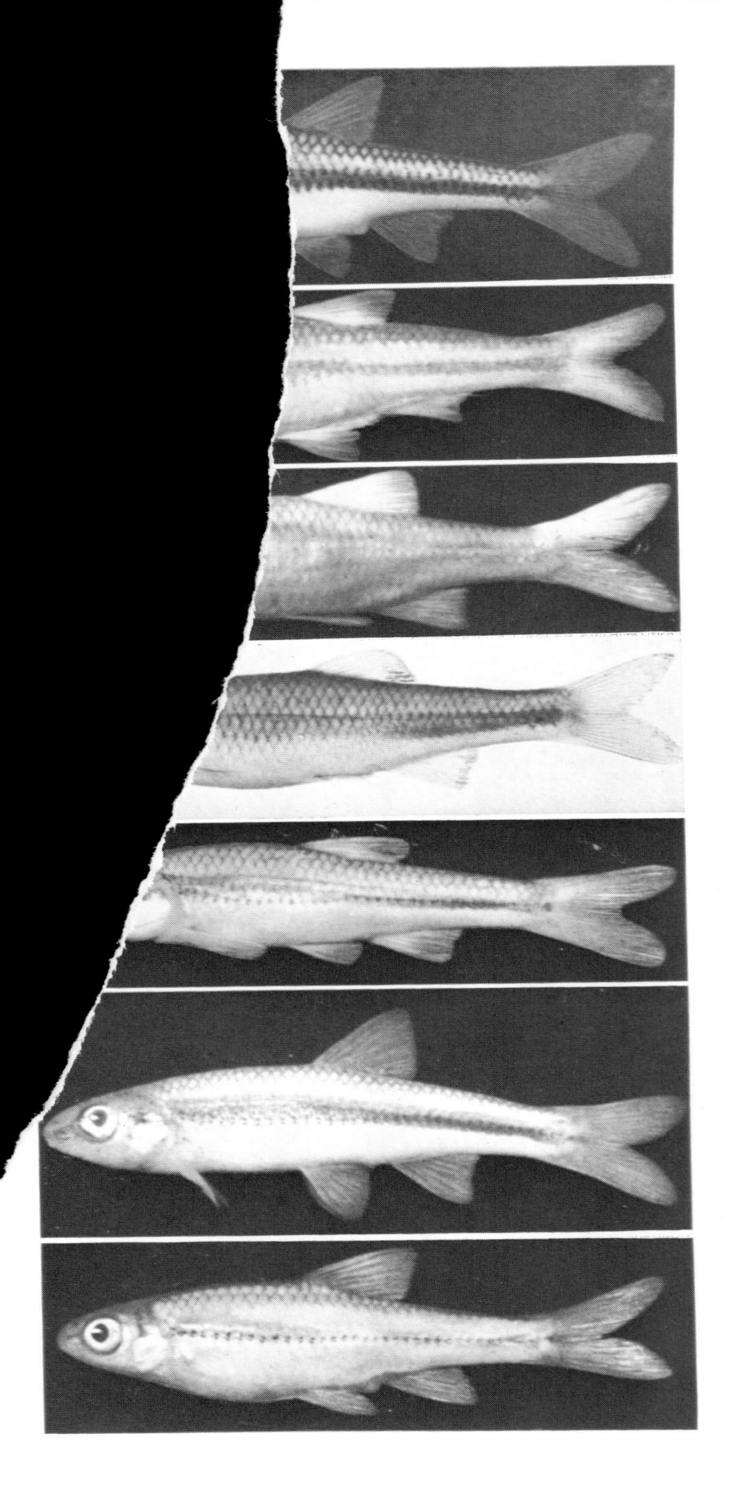

Fig. 147.
Blackchin shiner
Notropis heterodon (Cope)
Adult
About 1.4 × natural size
Mich. (F. N. B.)

Fig. 148.
Bigeye shiner
Notropis boops Gilbert
Adult male
About 1.5 × natural size
Okla., Sequoyah Co.

Fig. 149.
Satinfin shiner
Notropis analostanus (Girard)
Adult male
About 1.1 × natural size
N. Y., Tompkins Co.

Fig. 150.
Spotfin shiner
Notropis spilopterus (Cope)
Adult
About natural size
Mich. (I. F. R.)

Fig. 151.
Central bigmouth shiner
Notropis dorsalis dorsalis (Agassiz)
Adult female
About 1.5 × natural size
Mich., Allegan Co.

Fig. 152.
Northeastern sand shiner
Notropis deliciosus stramineus (Cope)
Adult (see also col. pl. 16)
About 1.2 × natural size
Mich. (F. N. B.)

Fig. 153.
Northern swallowtail shiner
Notropis procne procne (Cope)
Adult female
About 1.2 × natural size
Va.

Fig. 154.
Northern mimic shiner
Notropis volucellus volucellus (Cope)
Adult
About 1.2 × natural size
Mich. (F. N. B.)

Fig. 155.
Northern blacknose shiner
Notropis heterolepis heterolepis Eigenmann and Eigenmann
Adult
About 1.3 × natural size
Mich. (F. N. B.)

Fig. 156.
Bridled shiner
Notropis bifrenatus (Cope)
Adult male
About 1.5 × natural size
N. J., Passaic Co.

Fig. 157.
Pugnose shiner
Notropis anogenus Forbes
Ripe female
About 1.9 × natural size
Ohio, Ottawa Co.

Fig. 158.
Suckermouth minnow
Phenacobius mirabilis (Girard)
Adult female (see also col. pl. 17)
About 0.9 × natural size
Ind., Huntington Co.

Fig. 159.
Silverjaw minnow
Ericymba buccata Cope
Adult
About 1.3 × natural size
Mich., Monroe Co.

Fig. 160.
Brassy minnow
Hybognathus hankinsoni Hubbs
Adult male (dark flecks due to parasites)
About 1.1 × natural size
Mich., Huron Co.

Fig. 161.
Eastern silvery minnow
Hybognathus nuchalis regia Girard
Adult male
About 1.5 × natural size
N. Y., Madison Co.

Fig. 162.
Northern fathead minnow
Pimephales promelas promelas Rafinesque
Mature female
About 1.6 × natural size
Mich. (I. F. R.)

Fig. 163.
Northern fathead minnow
Pimephales promelas promelas Rafinesque
Breeding male (note nuptial tubercles on snout; not in very high color, see also col. pl. 18)
About 1.3 × natural size
Mich. (I. F. R.)

Fig. 164.
Northern bullhead minnow
Pimephales vigilax perspicuus (Girard)
Adult male
About 1.5 × natural size
Wis., Crawford Co.

Fig. 165.
Bluntnose minnow
Pimephales notatus (Rafinesque)
Adult female (with back view to show small predorsal scales)
About 1.2 × natural size
Mich. (F. N. B.)

Fig. 166.
Bluntnose minnow
Pimephales notatus (Rafinesque)
Breeding male
About 1.1 × natural size
Mich. (I. F. R.)

Fig. 167.
Central stoneroller
Campostoma anomalum pullum (Agassiz)
Gravid female
About 0.7 × natural size
Mich. (I. F. R.)

Fig. 168.
Central stoneroller
Campostoma anomalum pullum (Agassiz)
Breeding male (see also col. pl. 19)
About 0.7 × natural size
Mich. (I. F. R.)

NORTH AMERICAN CATFISH FAMILY —ICTALURIDAE

Fig. 169.
Northern channel catfish
Ictalurus punctatus punctatus (Rafinesque)
Adult (see also col. pl. 20)
About 0.2 × natural size
Mich., L. St. Clair (I. F. R.)

Fig. 170.
Northern black bullhead
Ictalurus melas melas (Rafinesque)
Adult (see also col. pl. 21)
About 0.3 × natural size
Ill. (I. N. H. S.)

Fig. 171.
Northern brown bullhead
Ictalurus nebulosus nebulosus (LeSueur)
Adult male
About 0.4 × natural size
Mich., Schoolcraft Co.

Fig. 172.
Northern yellow bullhead
Ictalurus natalis natalis (LeSueur)
Adult
About 0.4 × natural size
Mich. (F. N. B.)

Fig. 173.
Flathead catfish
Pylodictis olivaris (Rafinesque)
Immature (?) male
About 0.3 × natural size
N. C.

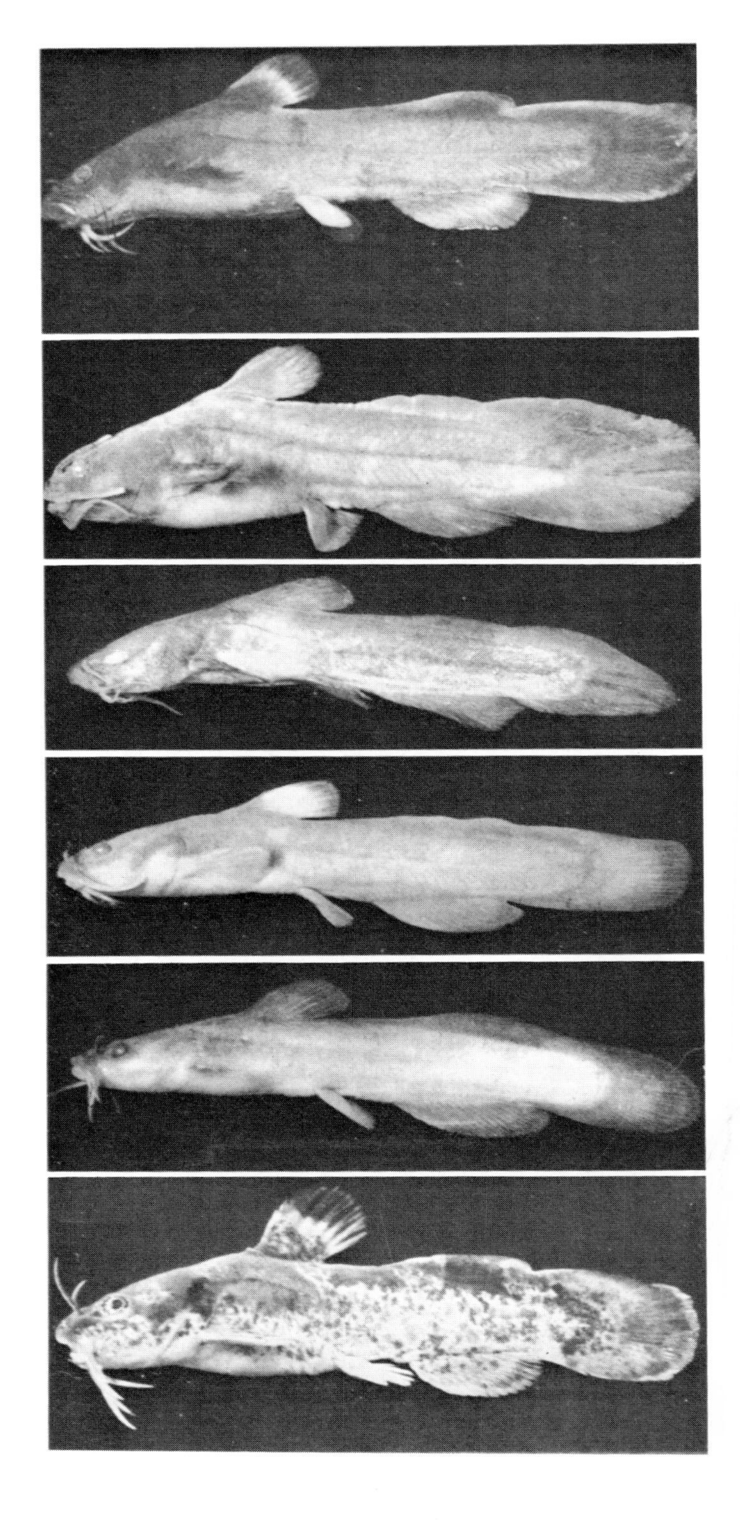

Fig. 174.
Stonecat
Noturus flavus Rafinesque
Adult
About 0.8 × natural size
Mich. (F. N. B.)

Fig. 175.
Tadpole madtom
Schilbeodes gyrinus (Mitchill)
Adult female
About 1.2 × natural size
Mich., Jackson Co.

Fig. 176.
Freckled madtom
Schilbeodes nocturnus (Jordan and Gilbert)
Adult male
About 0.9 × natural size
Mo., Dunklin Co.
Now thought not to occur in the Great Lakes basin (see text)

Fig. 177.
Common eastern madtom
Schilbeodes insignis insignis Richardson
Adult male
About 1.3 × natural size
N. Y., Broome Co.

Fig. 178.
Slender madtom
Schilbeodes exilis (Nelson)
Adult male
About 1.4 × natural size
Ark., Washington Co.

Fig. 179.
Brindled madtom
Schilbeodes miurus (Jordan)
Adult
About natural size
Mich. (F. N. B.)

Fig. 180.

Furious madtom
Schilbeodes furiosus (Jordan and Meek)
Adult male
About 0.8 × natural size
Mich., Washtenaw Co.

MUDMINNOW FAMILY—Umbridae

Fig. 181.

Central mudminnow
Umbra limi (Kirtland)
Adult male (see also col. pl. 23)
About 1.5 × natural size
Mich. (F. N. B.)

PIKE FAMILY—Esocidae

Fig. 182.

Western grass pickerel
Esox americanus vermiculatus LeSueur
Adult (see also col. pl. 24). About 0.4 × natural size
Mich. (F. N. B.)

Fig. 183.

Chain pickerel
Esox niger LeSueur
Immature male. About 0.4 × natural size
N. H., Grafton Co.

Fig. 184.

Chain pickerel
Esox niger LeSueur
Adult (note chain-like color pattern)
About 0.2 × natural size
(C. N. H. M.)

Fig. 185.

Northern pike
Esox lucius Linnaeus
Fingerling (note color pattern)
About 0.5 × natural size
Mich. (F. N. B.)

Fig. 186.

Northern pike
Esox lucius Linnaeus
Large young (with adult color pattern)
About 0.25 × natural size
Mich. (I. F. R.)

Fig. 187.
Great Lakes muskellunge
Esox masquinongy masquinongy Mitchill
Juvenile
About 0.3 × natural size
N. Y., Niagara R. (I. F. R.)

Fig. 188.
Great Lakes muskellunge
Esox masquinongy masquinongy Mitchill
Adult
About 0.1 × natural size
(C. N. H. M.)

FRESHWATER EEL FAMILY—ANGUILLIDAE

Fig. 189.
American eel
Anguilla bostoniensis (LeSueur)
Adult (see also col. pl. 25)
About 0.2 × natural size
Ill. (I. N. H. S.)

KILLIFISH FAMILY—CYPRINODONTIDAE

Fig. 190.
Eastern banded killifish
Fundulus diaphanus diaphanus (LeSueur)
Adult male
About 1.1 × natural size
N. H., Strafford Co.

Fig. 191.
Western banded killifish
Fundulus diaphanus menona Jordan and Copeland
Adult (with dorsal view to show markings on back)
(See also col. pl. 26)
About 1.2 × natural size
Mich. (F. N. B.)

Fig. 192.
Northern starhead topminnow
Fundulus nottii dispar (Agassiz)
Adult male
About 1.9 × natural size
Mo., Butler Co.

Fig. 193.
Blackstripe topminnow
Fundulus notatus (Rafinesque)
Adult (see also col. pl. 27)
About 1.3 × natural size
Mich. (F. N. B.)

LIVEBEARER FAMILY—POECILIIDAE

Fig. 194.

Western common gambusia
Gambusia affinis affinis (Baird and Girard)
Gravid female. About 1.7 × natural size
Ariz.

Fig. 195.

Western common gambusia
Gambusia affinis affinis (Baird and Girard)
Adult male (note modification of anal fin into a gonopodium)
About 2.7 × natural size
Ariz.

COD FAMILY—GADIDAE

Fig. 196.

American burbot
Lota lota lacustris (Walbaum)
Small adult. About 0.2 × natural size
N. Y. (N. Y. C. D.)

TROUTPERCH FAMILY—PERCOPSIDAE

Fig. 197.

Troutperch
Percopsis omiscomaycus (Walbaum)
Adult. About natural size
Mich. (F. N. B.)

PIRATEPERCH FAMILY—APHREDODERIDAE

Fig. 198.

Western pirateperch
Aphredoderus sayanus gibbosus LeSueur
Adult male. About natural size
Mich., Arenac Co.

BASS FAMILY—SERRANIDAE

Fig. 199.

White bass
Roccus chrysops (Rafinesque)
Immature female
About 0.2 × natural size
N. Y. (N. Y. C. D.)

Fig. 200.

Yellow bass
Morone interrupta Gill
Adult
About 0.5 × natural size
Ill. (I. N. H. S.)

PERCH FAMILY—PERCIDAE

Fig. 201.
Yellow perch
Perca flavescens (Mitchill)
Half-grown (for adult see col. pl. 28)
About 0.7 × natural size
Mich. (F. N. B.)

Fig. 202.
Sauger
Stizostedion canadense (Smith)
Half-grown male
About 0.3 × natural size
N. Y. (N. Y. C. D.)

Fig. 203.
Yellow walleye
Stizostedion vitreum vitreum (Mitchill)
Adult (see also col. pl. 29)
About 0.2 × natural size
Ill. (I. N. H. S.)

Fig. 204.
Blue walleye
Stizostedion vitreum glaucum Hubbs
Half-grown male
About 0.3 × natural size
N .Y. (N. Y. C. D.)

Fig. 205.
River darter
Hadropterus shumardi (Girard)
Adult male
About 1.7 × natural size
Mich., Iosco Co.

Fig. 206.
Blackside darter
Hadropterus maculatus (Girard)
Adult male (for adult female see col. pl. 30)
About natural size
Mich. (F. N. B.)

Fig. 207.
Slenderhead darter
Hadropterus phoxocephalus (Nelson)
Adult male
About 1.2 × natural size
Okla., Kay Co.

Fig. 208.

Gilt darter
Hadropterus evides (Jordan and Copeland)
Adult female
About 1.2 × natural size
Ind., Tippecanoe R.

Fig. 209.

Channel darter
Hadropterus copelandi (Jordan)
Adult male. About 1.7 × natural size
Mich., Monroe Co.

Fig. 210.

Northern logperch
Percina caprodes semifasciata (De Kay)
Adult (see also col. pl. 31)
About natural size
Mich. (F. N. B.)

Fig. 211.

Northern sand darter
Ammocrypta pellucida (Baird)
Adult male. About 1.5 × natural size
Quebec

Fig. 212

Central Johnny darter
Etheostoma nigrum nigrum Rafinesque
Adult female. About 1.9 × natural size
Mich. (F. N. B.)

Fig. 213.

Central Johnny darter
Etheostoma nigrum nigrum Rafinesque
Breeding male (note dark color)
About 1.6 × natural size
Mich. (F. N. B.)

Fig. 214.

Bluntnose darter
Etheostoma chlorosomum (Hay)
Adult male
About 1.9 × natural size
Miss., Wilkinson Co.

Fig. 215.

Bluebreast darter
Etheostoma camurum Cope
Adult female
About 1.2 × natural size
N. C., Swain Co.
Now thought not to occur in the Great Lakes basin (see text)

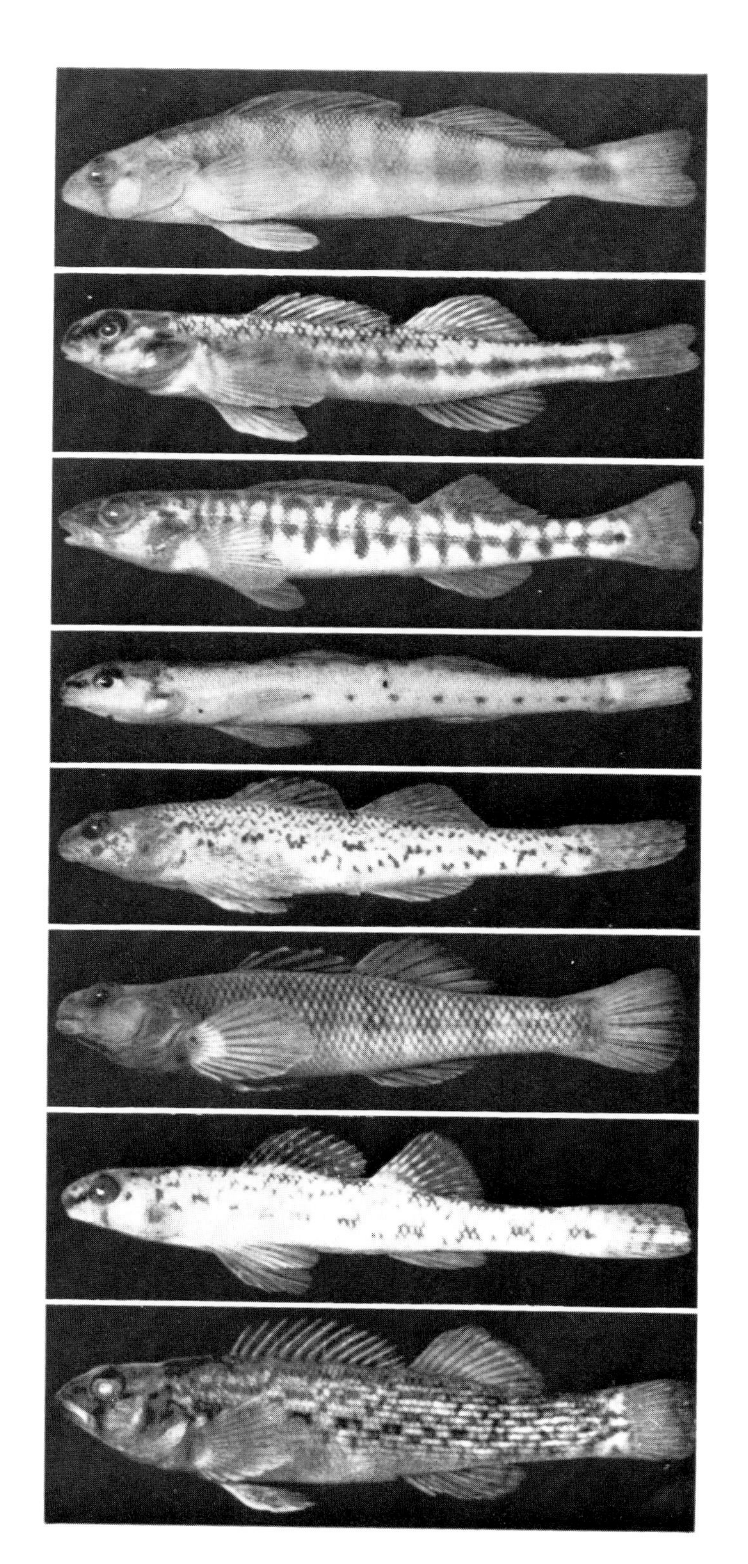

Fig. 216.
Eastern banded darter
Etheostoma zonale zonale (Cope)
Adult male
About 1.6 × natural size
Wis., Waupaca Co.

Fig. 217.
Mud darter
Etheostoma asprigene (Forbes)
Adult female
About 1.8 × natural size
Ill., L. Senachwine
Now thought not to occur in the Great Lakes basin (see text)

Fig. 218.
Iowa darter
Etheostoma exile (Girard)
Adult female (for adult male see col. pl. 32)
About 1.4 × natural size
Mich. (F. N. B.)

Fig. 219.
Rainbow darter
Etheostoma caeruleum Storer
Adult female
About 1.6 × natural size
Mich. (F. N. B.)

Fig. 220.
Rainbow darter
Etheostoma caeruleum Storer
Breeding male
About 1.5 × natural size
Mich. (F. N. B.)

Fig. 221.
Northern orangethroat darter
Etheostoma spectabile spectabile (Agassiz)
Adult male
About 1.7 × natural size
Ohio, Defiance Co.

Fig. 222.
Barred fantail
Etheostoma flabellare flabellare Rafinesque
Adult female
About 1.7 × natural size
Mich., Washtenaw Co.
(For male of barred fantail see next fig. and for male of striped fantail see col. pl. 33)

Fig. 223.
Barred fantail
Etheostoma flabellare flabellare Rafinesque
Adult male (see also col. pl. 33)
About 1.9 × natural size
Mich., Washtenaw Co.

Fig. 224.
Least darter
Etheostoma microperca Jordan and Gilbert
Mature female
About 3.0 × natural size
Mich. (F. N. B.)

Fig. 225.
Least darter
Etheostoma microperca Jordan and Gilbert
Adult male (with ventral view to show enlarged pelvics)
About 3.0 × natural size
Mich. (F. N. B.)

Fig. 226.
Northern greenside darter
Etheostoma blennioides blennioides Rafinesque
Adult female
About 0.8 × natural size
Mich. (F. N. B.)

Fig. 227.
Northern greenside darter
Etheostoma blennioides blennioides Rafinesque
Adult male (see also col. pl. 34)
About 0.7 × natural size
Mich. (F. N. B.)

SUNFISH FAMILY—CENTRARCHIDAE

Fig. 228.
Northern smallmouth bass
Micropterus dolomieui dolomieui Lacépède
Fingerling
About natural size
Mich. (F. N. B.)

Fig. 229.
Northern smallmouth bass
Micropterus dolomieui dolomieui Lacépède
Adult female (see also col. pl. 35)
About 0.22 × natural size
N. Y. (N. Y. C. D.)

Fig. 230.
Northern largemouth bass
Micropterus salmoides salmoides (Lacépède)
Juvenile
About 0.4 × natural size
Mich. (F. N. B.)

Fig. 231.
Northern largemouth bass
Micropterus salmoides salmoides (Lacépède)
Adult female (see also col. pl. 36)
About 0.2 × natural size
N. Y. (N. Y. C. D.)

Fig. 232.
Warmouth
Chaenobryttus gulosus (Cuvier)
Half-grown
About natural size
Mich. (F. N. B.)

Fig. 233.
Green sunfish
Lepomis cyanellus Rafinesque
Adult (see also col. pl. 37)
About 0.8 × natural size
Mich. (F. N. B.)

Fig. 234.
Pumpkinseed
Lepomis gibbosus (Linnaeus)
Juvenile
About 1.1 × natural size
Mich. (F. N. B.)

Fig. 235.
Pumpkinseed
Lepomis gibbosus (Linnaeus)
Adult female
About 0.4 × natural size
Ont., St. Lawrence R.

Fig. 236
Common bluegill
Lepomis macrochirus macrochirus Rafinesque
Half-grown (for adult male see col. pl. 38)
About natural size
Mich. (F. N. B.)

Fig. 237.
Orangespotted sunfish
Lepomis humilis (Girard)
Adult female
About 1.4 × natural size
S. D., Bon Homme Co.

Fig. 238.
Orangespotted sunfish
Lepomis humilis (Girard)
Adult male
About 1.4 × natural size
S. D., Bon Homme Co.

Fig. 239.
Northern longear sunfish
Lepomis megalotis peltastes Cope
Adult male (margin of ear flap bright red in life)
About natural size
Mich. (F. N. B.)

Fig. 240.
Northern rock bass
Ambloplites rupestris rupestris (Rafinesque)
Half-grown (for adult see col. pl. 39)
About natural size
Mich. (F. N. B.)

Fig. 241.
White crappie
Pomoxis annularis Rafinesque
Juvenile (for adult see col. pl. 40)
About natural size
Mich. (F. N. B.)

Fig. 242.

Black crappie
Pomoxis nigromaculatus (LeSueur)
Mature male (see also col. pl. 41)
About 0.3 × natural size
Mich., Washtenaw Co. (J. V. C.)

SILVERSIDE FAMILY—Atherinidae

Fig. 243.

Northern brook silverside
Labidesthes sicculus sicculus (Cope)
Adult (see also col. pl. 42)
About natural size
Mich. (F. N. B.)

DRUM FAMILY—Sciaenidae

Fig. 244.

Freshwater drum
Aplodinotus grunniens Rafinesque
Adult (see also col. pl. 43)
About 0.2 × natural size
Ill. (I. N. H. S.)

SCULPIN FAMILY—Cottidae

Fig. 245.

Great Lakes fourhorned sculpin
Myoxocephalus quadricornis thompsonii (Girard)
Adult female
About 0.9 × natural size
Mich., L. Huron

Fig. 246.

Spoonhead sculpin
Cottus ricei (Nelson)
Adult
About 1.2 × natural size
Mich. (F. N. B.)

Fig. 247.
Northern mottled sculpin
Cottus bairdii bairdii Girard
Adult
About 1.2 × natural size
Mich. (F .N. B.)

Fig. 248.
Eastern slimy sculpin
Cottus cognatus gracilis Heckel
Adult male
About 1.2 × natural size
Mich., Antrim Co.

STICKLEBACK FAMILY—GASTEROSTEIDAE

Fig. 249.
Brook stickleback
Eucalia inconstans (Kirtland)
Adult (see also col. pl. 44)
About 1.5 × natural size
Mich. (F. N. B.)

Fig. 250.
Ninespine stickleback
Pungitius pungitius (Linnaeus)
Adult male
About 1.4 × natural size
Mich., Macomb Co.

Fig. 251.
Threespine stickleback
Gasterosteus aculeatus Linnaeus
Adult female
About 1.9 × natural size
Me., Penobscot Co.

Index

INDEX

Figures preceded by an asterisk indicate principal page reference.

G

H

I

J

K

L

M

N

O

P

Q

R

S

T

U

V

W

X

Z

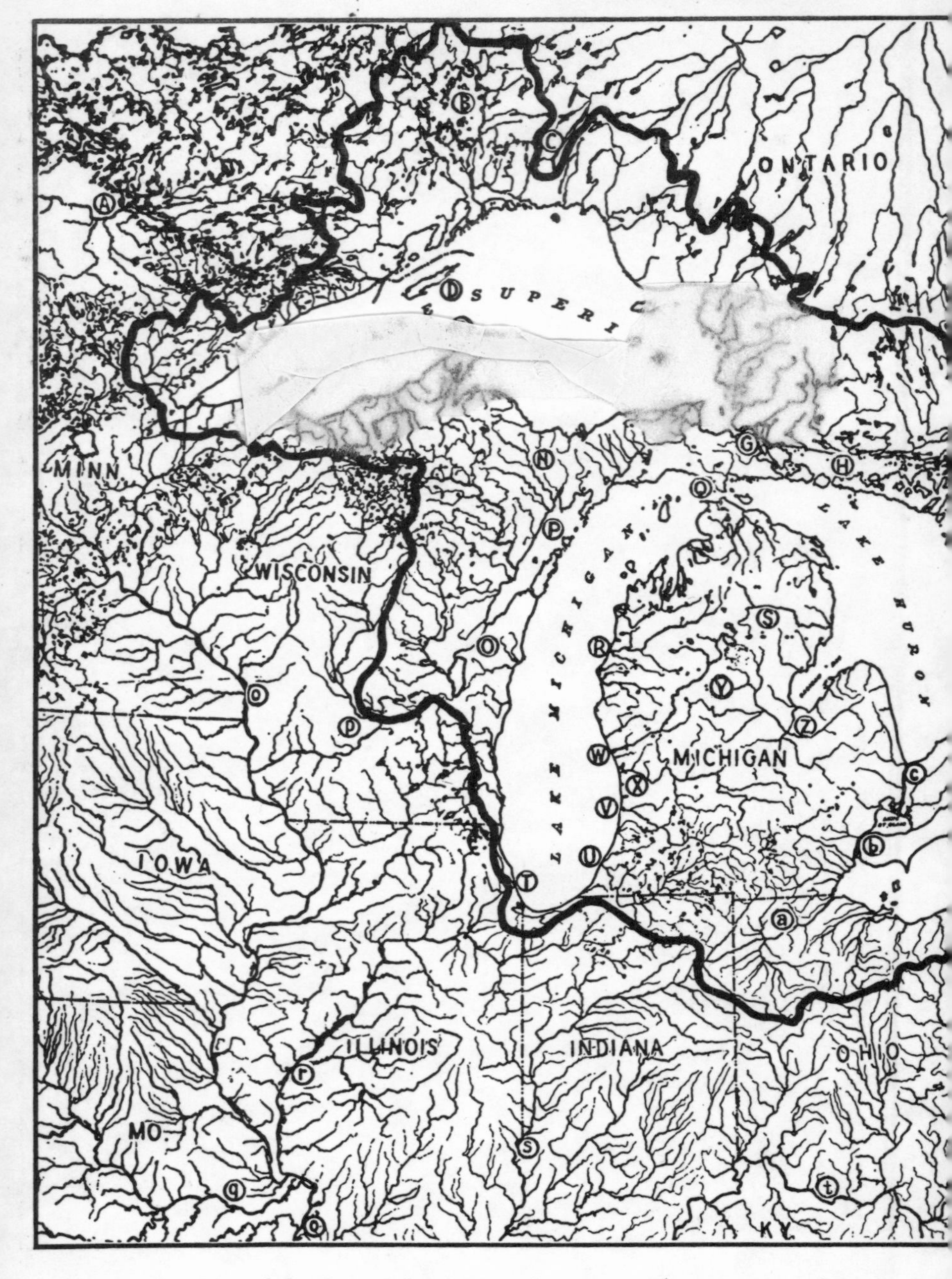

Map of the Great Lakes basin and environs indicating scope of this book a

A. Rainy River
B. Lake Nipigon
C. Long Lake
D. Isle Royale
E. Keweenaw Peninsula
F. Hudson Bay drainage
G. St. Mary's River
H. North Channel
I. Lake Abitibi
J. Lake Nipissing
K. Lake Simcoe

L. Ottawa River
M. St. Lawrence River
N. Upper Peninsula of Michigan
O. Fox River, Wisconsin
P. Green Bay
Q. Straits of Mackinac
R. Manistee River, Michigan
S. Au Sable River, Michigan
T. Chicago River and D

U. St. Joseph River, Mich and Indiana
V. Kalamazoo River, Michigan
W. Muskegon River, Michigan
X. Grand River, Michigan
Y. Lower Peninsula, Michigan
Z. Saginaw River, Michig